ENERGY SCIENCE, ENGINEERING AND TECHNOLOGY

U.S. PETROLEUM REFINING INDUSTRY AND THE GLOBALIZATION OF PETROLEUM PRODUCTS

ENERGY SCIENCE, ENGINEERING AND TECHNOLOGY

Additional books in this series can be found on Nova's website under the Series tab.

Additional e-books in this series can be found on Nova's website under the e-book tab.

ENERGY SCIENCE, ENGINEERING AND TECHNOLOGY

U.S. PETROLEUM REFINING INDUSTRY AND THE GLOBALIZATION OF PETROLEUM PRODUCTS

COLLIN T. ROBERTS
EDITOR

nova publishers
New York

Library of Congress Cataloging-in-Publication Data

ISBN: 978-1-63321-472-9

Published by Nova Science Publishers, Inc. † New York

CONTENTS

PREFACE

The U.S. petroleum refining industry—the largest refining industry in the world—experienced a period of high product prices and industry profits from the early 2000s through about 2007. Since the recession of 2007 to 2009, the industry has been in transition. Federal and state agencies regulate petroleum refining and the use of petroleum products to protect human health and the environment, as well as for other purposes. EPA, DOT, and California strengthened five key regulations, including EPA and DOT's coordinated fuel economy and GHG vehicle emission standards, and EPA's RFS, which has required that refiners and others ensure transportation fuels include increasing amounts of renewable fuels such as ethanol produced from corn. This book examines major changes that have recently affected the industry and the future of the industry.

Chapter 1 – The U.S. petroleum refining industry— the largest refining industry in the world—experienced a period of high product prices and industry profits from the early 2000s through about 2007. Since the recession of 2007 to 2009, the industry has been in transition. Federal and state agencies regulate petroleum refining and the use of petroleum products to protect human health and the environment, as well as for other purposes. EPA, DOT, and California recently proposed or strengthened five key regulations, including EPA and DOT's coordinated fuel economy and GHG vehicle emission standards, and EPA's RFS, which has required that refiners and others ensure transportation fuels include increasing amounts of renewable fuels such as ethanol produced from corn. GAO was asked to provide information on the domestic petroleum refining industry. This report examines: (1) major changes that have recently affected the industry and (2) the future of the industry. GAO reviewed information including studies by agencies and consultants and company financial filings; interviewed stakeholders, including agency officials and representatives of refiners and environmental organizations; and reviewed forecasts by the Energy Information Administration and others.

Chapter 2 – To better understand how changes in domestic and international petroleum products markets have affected prices, GAO was asked to evaluate trends in (1) the international trade of petroleum products, (2) refining capacity and intensity of refining capacity use internationally and in the United States, (3) international and domestic crude oil and petroleum product inventories, and (4) domestic petroleum supply infrastructure. To address these objectives, the authors reviewed numerous studies, evaluated data, and spoke to many industry officials and experts and agency officials.

In: U.S. Petroleum Refining Industry … ISBN: 978-1-63321-472-9
Editor: Collin T. Roberts

Chapter 1

PETROLEUM REFINING: INDUSTRY'S OUTLOOK DEPENDS ON MARKET CHANGES AND KEY ENVIRONMENTAL REGULATIONS*

United States Government Accountability Office

WHY GAO DID THIS STUDY

The U.S. petroleum refining industry— the largest refining industry in the world—experienced a period of high product prices and industry profits from the early 2000s through about 2007. Since the recession of 2007 to 2009, the industry has been in transition.

Federal and state agencies regulate petroleum refining and the use of petroleum products to protect human health and the environment, as well as for other purposes. EPA, DOT, and California recently proposed or strengthened five key regulations, including EPA and DOT's coordinated fuel economy and GHG vehicle emission standards, and EPA's RFS, which has required that refiners and others ensure transportation fuels include increasing amounts of renewable fuels such as ethanol produced from corn.

GAO was asked to provide information on the domestic petroleum refining industry. This report examines: (1) major changes that have recently affected the industry and (2) the future of the industry. GAO reviewed

* This is an edited, reformatted and augmented version of the United States Government Accountability Office publication, GAO-14-249, dated March 2014.

information including studies by agencies and consultants and company financial filings; interviewed stakeholders, including agency officials and representatives of refiners and environmental organizations; and reviewed forecasts by the Energy Information Administration and others.

WHAT GAO RECOMMENDS

GAO recommends that EPA identify the underlying causes of delays in issuing RFS standards and implement a plan to issue RFS standards on time. EPA generally agreed with GAO's findings and recommendations.

WHAT GAO FOUND

Stakeholders GAO contacted and information reviewed by GAO identified the following three major changes that have recently affected the domestic petroleum refining industry:

- *Increased production.* U.S. and Canadian crude oil production have increased, leading to lower costs of crude oil for some refiners. After generally declining for decades, monthly U.S. crude oil production increased over 55 percent compared with average production in 2008.
- *Declining consumption.* Domestic consumption of petroleum products declined by 11 percent from 2005 through 2012, resulting in a smaller domestic market for refiners.
- *Key regulations.* Two key regulations—the Environmental Protection Agency's (EPA) and Department of Transportation's (DOT) coordinated fuel economy and greenhouse gas (GHG) vehicle emission standards, as well as EPA's Renewable Fuel Standard (RFS)—have contributed to declining petroleum-based fuel consumption. For some refiners, compliance with the RFS increased costs in the first half of 2013, though costs have since declined to some degree from their peak. According to some stakeholders GAO contacted, this was primarily due to RFS requirements exceeding the capability of the transportation fuel infrastructure to distribute and the fleet of vehicles to use renewable fuels. Moreover, EPA has missed the statutory deadline to issue regulations establishing annual RFS

blending standards since 2009. EPA has not systematically identified the underlying causes of these delays or changed its approach in order to avoid them. A late RFS contributes to industry uncertainty, which can increase costs because industry cannot plan and budget effectively, according to some stakeholders.

Stakeholders GAO contacted and information reviewed generally suggested that the U.S. refining industry's outlook depends on the following factors:

- *Domestic consumption.* Future consumption of petroleum products is uncertain, with projections ranging from stable to slightly increasing through 2020 but not returning to consumption levels of the past. Forecasts GAO reviewed suggest higher future refinery production in scenarios with higher domestic consumption.
- *Costs of key regulations.* The extent to which requirements in the key regulations increase costs for refiners will affect the industry's outlook. For example, future costs to comply with RFS may depend on the annual renewable fuel volumes EPA sets and whether EPA issues annual RFS standards on time. In general, increasing costs may be absorbed by refiners (i.e., by reducing their profits), be passed on to consumers through higher prices, or both.
- *Foreign markets.* The U.S. refining industry has increasingly relied on foreign markets. Exports grew from 7 percent of production in 2007 to 17 percent in 2012. The extent to which domestic refiners export their products will depend on the competitiveness of U.S. refiners. Factors that may affect competitiveness include domestic environmental regulations, levels of U.S. and Canadian crude oil production, and the balance between global refining capacity and demand for petroleum products.

ABBREVIATIONS

CAFE	Corporate Average Fuel Economy
CARB	California Air Resources Board
CI	carbon intensity
DOE	Department of Energy
DOT	Department of Transportation

EIA	Energy Information Administration
EPA	Environmental Protection Agency
GHG	greenhouse gas
IEA	International Energy Agency
LCFS	Low Carbon Fuel Standard
NHTSA	National Highway Traffic Safety Administration
OMB	Office of Management and Budget
ppm	parts per million
PSD	Prevention of Significant Deterioration
RFS	Renewable Fuel Standard
RIN	renewable identification number

March 14, 2014

The Honorable David Vitter
Ranking Member
Committee on Environment and Public Works
United States Senate

The Honorable James Inhofe
United States Senate

The petroleum refining industry in the United States—the largest refining industry in the world—experienced what some observers and industry analysts have called a "golden age" from the early 2000s through about 2007. As we reported in 2007, this period was characterized by increased petroleum product prices, as well as higher price volatility and industry profits.[1] Conditions changed with the recession of 2007 to 2009—the most severe in this country since the 1930s—and the industry has since been in transition. Amid declining refining profits, seven refineries have closed since 2008, and the remaining refineries have not been run as intensively. Conditions have improved more recently, with an Energy Information Administration (EIA) official noting that the industry may be in a second golden age.[2] Petroleum refineries process crude oil into transportation fuels and other products (e.g., gasoline, diesel, and kerosene) that together comprise the largest source of energy in the United States—36 percent of the energy consumed in 2012, according to data from EIA. Domestic petroleum refineries produce the majority of the fuels used in our transportation system, which also uses a small portion of other fuels such as renewable fuels produced from corn, sugar cane, and soybeans.

Federal and state agencies regulate aspects of petroleum refining and the use of refined petroleum products for several purposes, including to protect human health and the environment, and to respond to concerns regarding the nation's dependence on imported crude oil. Petroleum refining and the combustion of gasoline and diesel emit carbon dioxide, a greenhouse gas (GHG) linked to climate change. To address climate change and other concerns, the Environmental Protection Agency (EPA), the primary federal agency responsible for implementing many of the nation's environmental laws, along with the Department of Transportation (DOT) and the state of California recently proposed or strengthened five key environmental regulations as follows:[3]

- *Renewable Fuel Standard (RFS).* Established in light of concerns such as climate change and the nation's dependence on imported crude oil, the RFS provides that U.S. transportation fuels must contain certain percentages of renewable fuels. Under the law establishing the program, volumes of renewable fuels were to increase over time, though EPA determines requirements annually and may reduce percentages under certain circumstances.[4]
- *Corporate Average Fuel Economy (CAFE) and GHG vehicle emission standards.* A coordinated program in which DOT establishes fleet-wide gas mileage or fuel economy standards and EPA establishes GHG emission standards for vehicle manufacturers.
- *Tier 3 Motor Vehicle Emission and Fuel Standards.* EPA's standards for vehicles and transportation fuels to reduce emissions of certain air pollutants.
- *Stationary source GHG requirements.* EPA's new requirements to address GHG emissions at new refineries and refineries that undertake major modifications.
- *Low Carbon Fuel Standard (LCFS).* California's recently implemented LCFS requires reductions to the amount of carbon in California's transportation fuels. California is important for the domestic refining industry because it is the nation's largest consumer of petroleum-based transportation fuels. It also has the nation's third largest refining capacity, after Texas and Louisiana.

You asked us to provide information about the domestic petroleum refining industry and its market environment. This report examines what is known about: (1) major changes—including key environmental regulations—

that have recently affected the domestic petroleum refining industry and (2) major factors that may affect the future of the domestic petroleum refining industry—including its production, profitability, and competitiveness in foreign markets.

To conduct this work, we reviewed information, including studies by federal agencies and consultants, and company financial regulatory filings; and summarized the results of interviews with a nonprobability sample of 32 stakeholders. Stakeholders included representatives from refining companies, environmental organizations, consultants, and officials from federal and state agencies. We also visited several refineries of selected refining companies. We selected these stakeholders to represent broad and differing perspectives on these issues based on recommendations from agencies, industry associations, and others. We took into account such factors as the location and size of companies' refineries. Because we used a nonprobability sample, the views of stakeholders are not generalizable to all potential stakeholders but provide illustrative examples of the range of views. Similarly, the conditions at the refineries we visited are not generalizable to all refineries. The stakeholder views we summarize were not necessarily supported by all types of stakeholders, though we identify differing views where appropriate. The stakeholders we contacted and information we reviewed identified a number of changes that have affected industry and factors that may affect its future, and we report on those that were most often cited. Based on our research and information from stakeholders, we identified five key regulations that were recently strengthened or proposed, though other regulations may also affect the industry. To illustrate major changes affecting industry over time, we summarized historical data from EIA. We took several steps to assess the reliability of EIA data, including reviewing documentation, interviewing EIA staff, and consulting with stakeholders. We determined the EIA data to be sufficiently reliable for the purposes of this report. To provide information about major factors that could affect the industry, we also reviewed forecasts from EIA, the International Energy Agency (IEA),[5] and IHS,[6] and summarized projections through 2020 under different scenarios. While forecasts are subject to inherent uncertainties, we found these forecasts to be reasonable for describing a range of views about potential conditions of the domestic refining industry and major factors that will help determine these conditions.[7] Appendix I provides additional information on our scope and methodology, and appendix II lists the stakeholders we interviewed.

We conducted this performance audit from November 2012 to March 2014 in accordance with generally accepted government auditing standards. Those standards require that we plan and perform the audit to obtain sufficient, appropriate evidence to provide a reasonable basis for our findings and conclusions based on our audit objectives. We believe that the evidence obtained provides a reasonable basis for our findings and conclusions based on our audit objectives.

BACKGROUND

This section describes the petroleum refining industry and the five key regulations that we reviewed.

Petroleum Refining Industry

According to data from EIA, there were 143 petroleum refineries in the United States as of January 2013, with a capacity to process 17.8 million barrels of crude oil per day.[8] While there are refineries in most regions of the country, most refining capacity (almost 90 percent) is located in the Gulf Coast, West Coast, and Midwest regions (see Figure 1).[9] These refineries employed over 70,000 people in 2013.

Refineries process crude oil into products primarily through a distillation process that separates crude oil into different fractions based on their boiling points, which can then be further processed into final products. One barrel of crude oil can be processed into varying amounts of gasoline, diesel, jet fuel, and other petroleum products depending on the configuration—or complexity—of the refinery and the type of crude oil that is being refined. Through the addition of specialized equipment, refineries can be optimized—or "upgraded"—to produce greater proportions of specific types of products or to use different types of crude oil. For example, a coker unit upgrades the low-value residual oil from the distillation process into higher value products such as diesel, increasing a refinery's ability to process heavier crude oils.[10] As shown in Figure 2, from a barrel of crude oil, U.S. refineries primarily produce gasoline, diesel, and jet fuel that are used in the transportation sector, along with heating oil and liquefied petroleum gases such as propane used in home heating.[11]

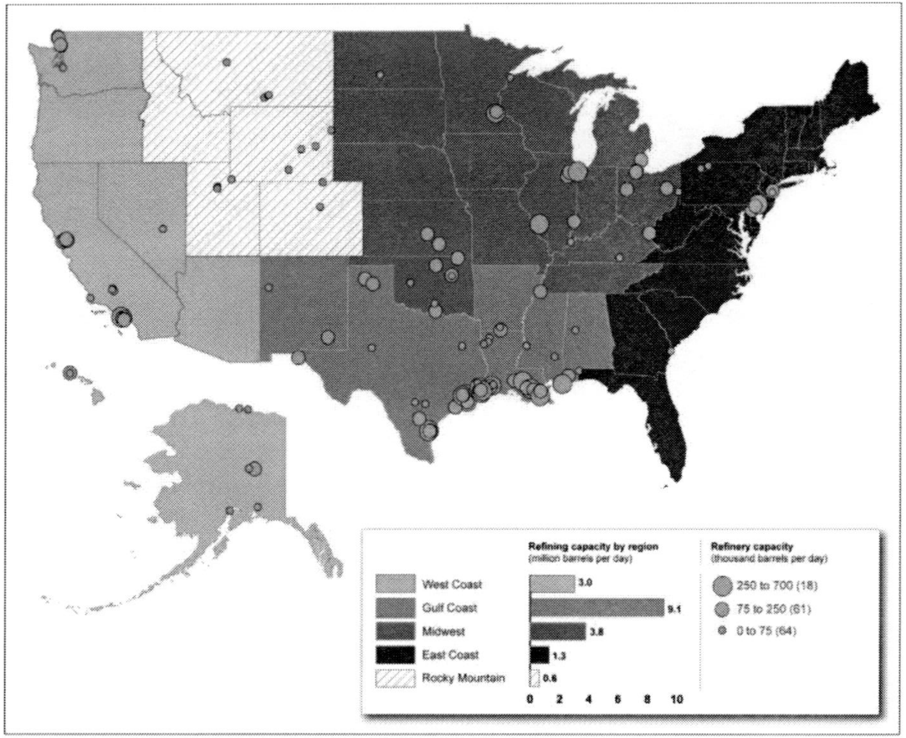

Sources: GAO analysis of EIA data; MapInfo (map).
Note: These regions are Petroleum Administration for Defense Districts.

Figure 1. Location and Capacity of Petroleum Refineries and Capacity by Region as of January 1, 2013.

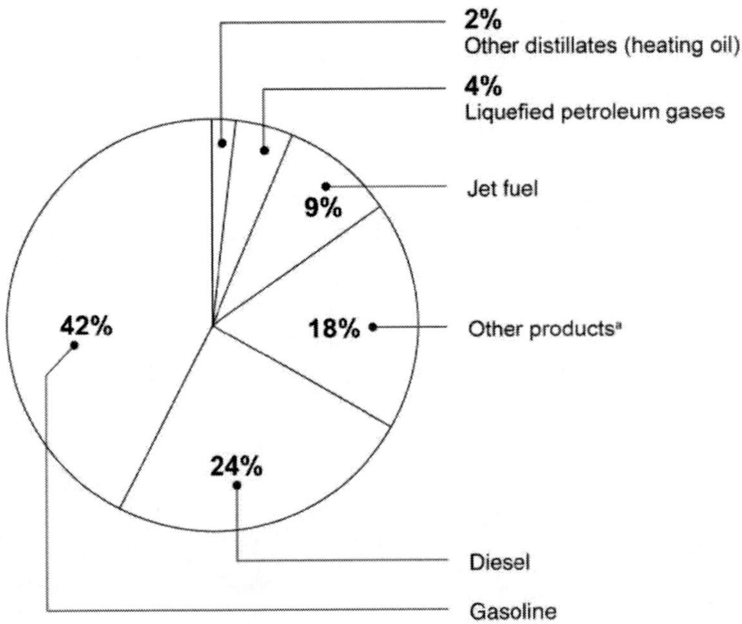

Source: GAO analysis of EIA data.

[a] Other products include heavy fuel oil, petroleum coke, asphalt and road oil, and other miscellaneous products.

Note: Numbers do not add to 100 due to independent rounding.

Figure 2. Petroleum Products Made from a Barrel of Crude Oil in the United States in 2012.

The U.S. petroleum refining industry consists of firms of varying sizes that, in addition to operating refineries, may also have operations in other related industry segments: (1) the upstream segment, which consists of the exploration for and production of crude oil; (2) the midstream segment, which consists of pipelines and other infrastructure used to transport crude oil and refined products; (3) the downstream segment, which consists of the refining and marketing of petroleum products such as gasoline and heating oil; and (4) the renewable fuels segment, where biorefineries produce renewable fuels that are blended with petroleum products at wholesale terminals before being distributed to consumers. To varying degrees, refiners may primarily operate refineries—these are called merchant refiners—or may be integrated, participating in various other related industry segments. HollyFrontier

Corporation is an example of a merchant refiner that purchases crude oil from unaffiliated producers and sells refined products to other companies operating retail fuel outlets, while Chevron is an example of a fully integrated company, a refiner that also produces crude oil and operates pipelines and retail fueling outlets across the United States.

Crude oil, petroleum products, and renewable fuels are transported between market participants through an extensive supply infrastructure including pipelines, tanker vessels, rail, trucks, wholesale terminals, and retail outlets. (See Figure 3) In 2012, refineries received the majority of their crude oil by pipeline (over 50 percent) and by tanker vessel (37 percent), with trucks and rail generally playing a more limited role according to EIA data.

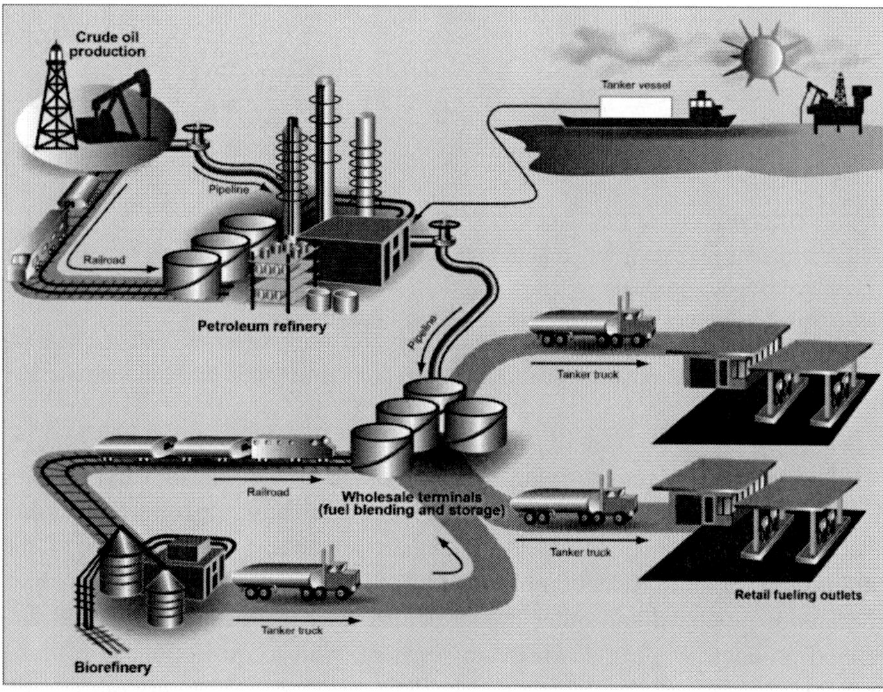

Source: GAO.

Note: Other means of transportation are also used to move petroleum and renewable fuels to wholesale terminals. For example, for ethanol, barges are also used to a limited extent.

Figure 3. Primary Transportation Infrastructure for Crude Oil, Petroleum Products, and Renewable Fuels.

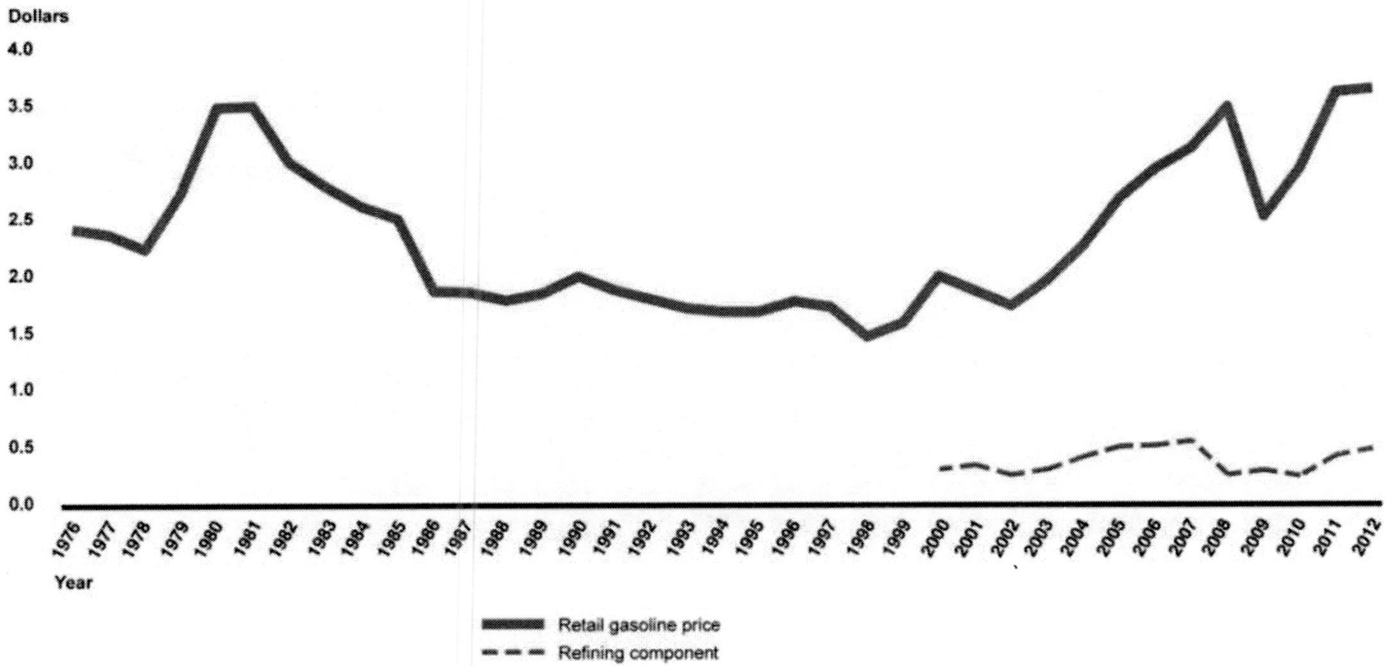

Source: GAO analysis of EIA data.

Note: EIA began reporting estimates of the components of retail prices in 2000. Prices were converted to 2012 dollars using the Bureau of Labor Statistics' Consumer Price Index.

Figure 4. U.S. Real Annual Average Retail Regular Gasoline Price and Refining Component of Retail Price, 1976-2012.

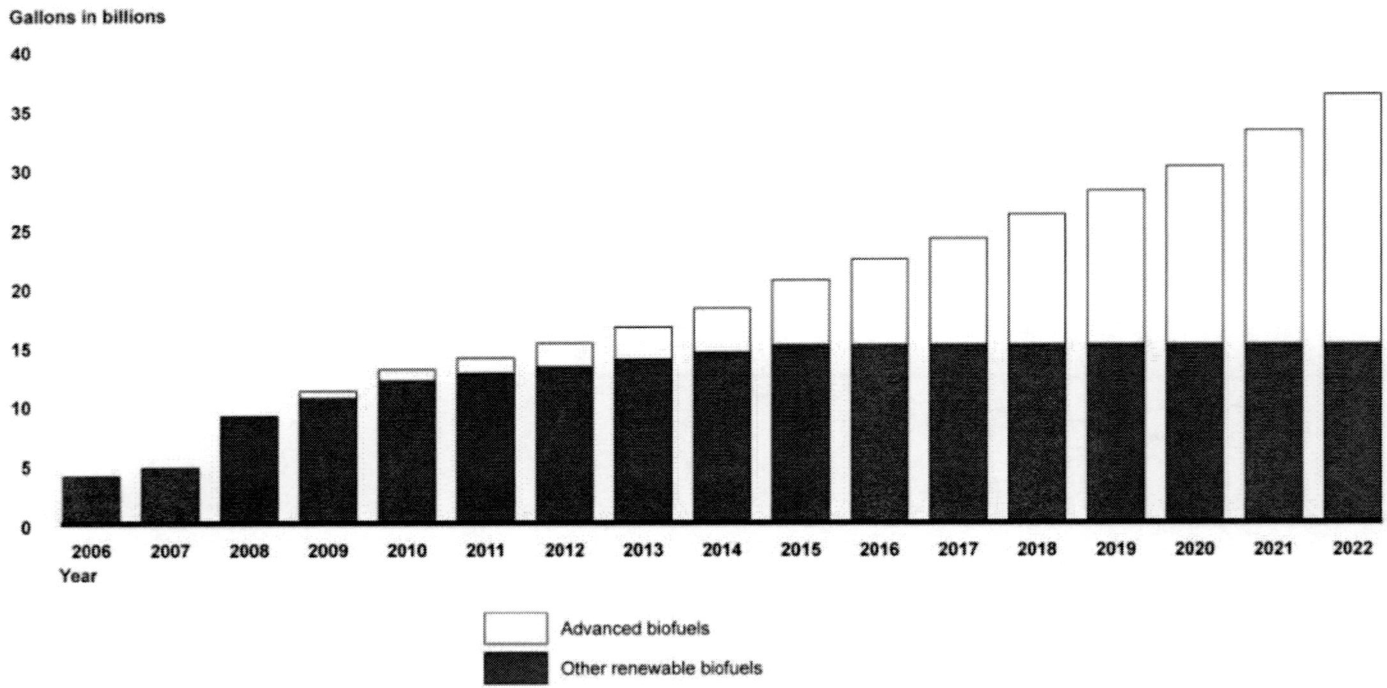

Source: Energy Independence and Security Act of 2007, Pub. L. No. 110-140 §202(a)(2) (2007).

Figure 5. Renewable Fuel Standard Volumes Established by the Energy Independence and Security Act of 2007.

As we reported in 2007, according to industry officials and experts, the refining industry was a low-return industry for much of the prior two decades.[12] Retail prices for regular gasoline averaged $3.63 per gallon in 2012, the highest annual average price when adjusted for inflation since 1976, the earliest comparable data available from EIA.[13] Retail prices have declined in 2013—gasoline averaged $3.55 in the first half of 2013 and $3.18 in November 2013—but are still near historic highs. Market dynamics anywhere along the supply chain can influence consumer prices, beginning with upstream crude oil production, the production of renewable fuels, through downstream refining and retailing.[14] According to EIA data, increases in crude oil costs have been the largest component of the recent increases in gasoline prices. The refining component of prices—including labor, materials, energy, and other costs of the refining process, as well as profits to refinery owners—has fluctuated over time but has not increased in a significant way since 2000, when EIA began reporting estimates of the components of retail prices (see Figure 4).

Key Regulations Affecting the Domestic Refining Industry

The five key environmental regulations affecting the domestic refining industry that we reviewed are concerned with various health, environmental, and other issues.

RFS

Under the RFS, since 2006, transportation fuels sold in the United States have been required to contain increasing amounts of renewable fuels such as ethanol and biodiesel. EPA is responsible for administering the RFS and annually issues regulations that establish the percentage of gasoline and diesel fuels that refiners, importers, and other obligated parties must ensure are renewable fuels. Congress established the RFS in light of concerns such as climate change and the nation's dependence on imported crude oil. As shown in figure 5, the law generally required that transportation fuels contain 9 billion gallons of renewable fuels in 2008, and that volumes increase 4-fold through 2022 to 36 billion gallons. The Administrator of EPA is authorized to waive the RFS levels established in the act if the Administrator determines—in consultation with the Secretaries of Agriculture and Energy—that implementing the requirement would severely harm the economy or environment, that there is an inadequate domestic supply, or in certain other situations.[15] The major source of renewable fuels has traditionally been ethanol produced from corn; however, as we reported in 2009, the increased cultivation of corn for ethanol, its conversion into renewable fuels, and the storage and use of these fuels could affect water supplies, water quality, air quality, soil quality, and biodiversity.[16] Under the RFS' statutory provisions, the increasing amounts of renewable fuels are to primarily come from renewable fuels other than corn ethanol—called advanced biofuels—that meet certain criteria, including reducing GHG emissions by at least 50 percent compared with the gasoline or diesel fuel they displace.[17] According to EPA, achieving the RFS' statutory blending levels in 2022 could result in total benefits—including those related to overall fuel costs, energy security, health, and GHG effects—of between $13 and $26 billion and could reduce GHG emissions by 138 million metric tons of carbon dioxide equivalent emissions, equal to taking about 27 million vehicles off the road.[18]

CAFE and GHG Vehicle Emission Standards

The federal government has regulated vehicle fuel economy through CAFE standards since 1978 and, more recently, aligned these standards with new GHG vehicle emission standards in a joint national program aimed at reducing oil consumption and GHG emissions from the transportation sector.[19] CAFE standards are administered by the National Highway Traffic Safety Administration (NHTSA) and require that vehicle manufacturers meet fleet-wide average fuel economy standards for vehicles.[20] The Energy Independence and Security Act of 2007 instituted several changes to the CAFE standards and, in 2009, the administration announced a new program to increase vehicle fuel economy and reduce vehicle GHG emissions, which was implemented by a joint rulemaking with NHTSA raising CAFE standards and EPA establishing the first GHG emissions standards for vehicles.[21] Although the CAFE and GHG vehicle emission standards are distinct, their targets were aligned for compliance purposes. NHTSA and EPA put the national program into place by issuing coordinated regulations covering vehicle model years 2012 to 2025.[22] As shown in figure 6, fuel economy standards for cars largely remained unchanged from 1990 through 2010, but vehicle manufacturers are now expected to meet increasingly stringent standards reaching the projected combined average fuel economy of about 50 miles per gallon by 2025— about 80 percent more efficient than required under the 2011 standards.[23] EPA estimated that the 2011-2025 standards may save consumers and businesses

$1.7 trillion, reduce oil consumption by 12 billion barrels, and reduce GHG emissions by 6 billion metric tons over the lifetime of the vehicles sold during model years 2011-2025.[24]

Tier 3 Motor Vehicle Emission and Fuel Standards

Under the Clean Air Act, EPA is authorized to establish certain standards for new motor vehicles and fuels to address air pollution that may reasonably be anticipated to endanger public health or welfare.[25] On May 21, 2013, EPA proposed the Tier 3 standards, and on March 3, 2014, EPA announced the final Tier 3 standards which establish more stringent vehicle emission standards and reduce the sulfur content of gasoline.[26] (Because EPA finalized this rulemaking after the draft report was completed and provided to agencies, the views of stakeholders and other information on Tier 3 that we reviewed and summarize in the rest of this report relate primarily to the proposed standards. EPA stated that the final rulemaking is very similar to the proposal, and that EPA made some changes—including to the sulfur provisions—based on public input.) According to EPA, more than 149 million Americans experience unhealthy levels of air pollution that has been linked to respiratory and cardiovascular problems and other adverse health effects. Cars and light trucks are significant contributors to air pollution, and EPA estimated that the Tier 3 standards will reduce pollution from such sources. The standards set more stringent tailpipe emission standards for new vehicles and generally require refiners to lower the sulfur content of gasoline from 30 parts per million (ppm) to 10 ppm on an annual average basis by 2017, among other things.[27] According to EPA, reducing the sulfur content of gasoline allows emissions control systems to work more effectively for existing and new vehicles and would therefore enable more stringent vehicle emissions standards. EPA estimated that the Tier 3 standards would reduce on-highway vehicle emissions of nitrogen oxides, a pollutant that has been liked to respiratory illnesses, by 10 percent in 2018, and 25 percent in 2030.[28] According to EPA estimates, by 2030, annual emission reductions from the Tier 3 standards would generate annual benefits of between $6.7 and $19 billion and prevent up to 2,000 premature deaths annually. EPA estimated that the vehicle and fuel standards would cost approximately $1.5 billion in 2030, including costs for refiners to install and operate equipment to remove sulfur from gasoline, as well as costs for vehicle manufacturers to improve the emissions performance of vehicles.

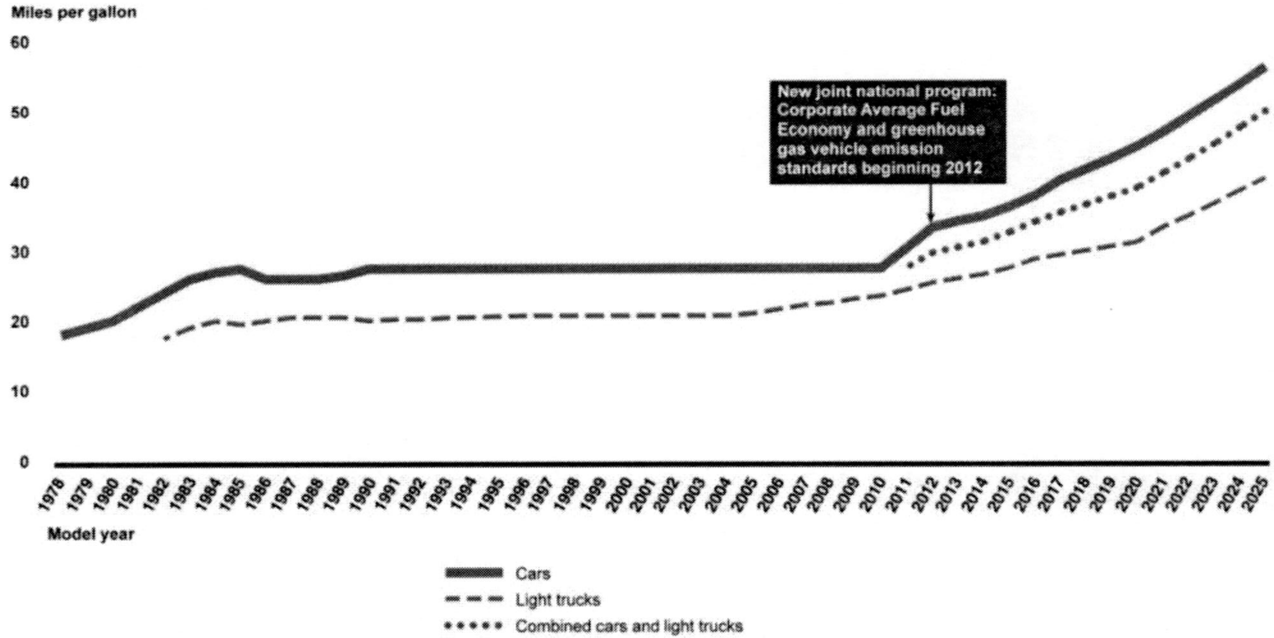

Source: GAO analysis of EPA and National Highway Traffic Safety Administration data.

Note: Data from 2011 on are projected fuel economy based on final standards, and data for 2022-2025 are based on projected standards. Projected fuel economy for 2017-2025 used model year 2008 as the baseline. Actual fuel economy may differ, depending on the fleet of vehicles sold and how manufacturers choose to meet the standards given flexibilities in the standards, including credits for improved air conditioning efficiency. The figure indicates CAFE standards, which are generally aligned with EPA's greenhouse gas vehicle emission standards.

Figure 6. Corporate Average Fuel Economy (CAFE) Standards, 1978-2025.

Stationary Source GHG Requirements

Under the Clean Air Act, EPA is authorized to take certain steps to address emissions from stationary sources, including refineries. EPA has regulated certain emissions of air pollutants from stationary sources for several decades, and EPA recently issued rules concerning how GHGs are to be included in certain existing permitting processes.[29] Specifically, permitting authorities[30] are to include GHG emission control requirements in Prevention of Significant Deterioration (PSD) permits and certain other permits issued to refineries and other stationary sources that trigger GHG emissions thresholds.[31] Since 2011, construction of any new refineries and certain refineries that are modified have generally been subject to the use of the "best available control technology" for GHG emissions. The best available control technology is determined for each facility based on an analysis of available technologies considering cost and other factors. According to EPA, in most cases, the best available control technology selected for GHGs are energy efficiency improvements. For example, for refineries, this could involve the installation of heat recovery units, which capture and use otherwise wasted heat in the refinery process. Such energy efficiency improvements can lower GHG emissions and other pollutants while reducing fuel consumption and saving money. Current regulations do not require existing facilities to take any steps to control GHG emissions unless they undertake a major modification. Examples of major modifications at a refinery include a significant expansion of crude oil processing units, or installing new secondary processing units that would increase GHG emissions above specified thresholds.

LCFS

California's LCFS aims to lower GHG emissions by reducing the level of carbon in transportation fuels. Established by the California Air Resources Board (CARB) following state legislation and an executive order, the LCFS has been fully in effect since January 2011. The LCFS would change the mix of fuels and vehicles in California to reduce emissions throughout the fuel "life cycle"—which includes emissions associated with producing, transporting, distributing, and using the fuel. To reduce emissions, carbon intensity (CI) scores are used, which are to reflect each fuel's life cycle GHG emissions. Refiners generally are required to ensure that the overall CI score for their fuels—which can include gasoline, diesel, and their blendstocks and substitutes—meets the annual carbon intensity target for a given year. Unlike the RFS, which requires certain types of renewable fuels be used, under LCFS refiners can meet the CI reduction targets using a variety of low carbon fuel technologies. Low carbon fuel technologies include renewable fuels from waste and cellulosic materials, natural gas, electricity used in plug-in vehicles, and hydrogen used in fuel cell vehicles. The original LCFS statewide reduction targets for gasoline, diesel, and their substitutes started at 0.25 percent of 2010 values in 2011 and increased to 10 percent by 2020. However, in 2013 a state Court of Appeal found that CARB must correct certain aspects of the procedures by which the LCFS was originally adopted.[32] CARB officials subsequently announced a regulatory package would be proposed in 2014, and that the 2013 standards—a 1 percent decrease in carbon intensity from 2010 values—will remain in effect through 2014.

To comply with LCFS, refiners can produce their own low carbon fuels, buy such fuels from other producers to blend into their products and sell on the market, or purchase credits generated by others. Refiners can also generate credits—which can be banked and traded—if their use of low carbon fuels results in greater-than-required carbon intensity reductions. CARB estimated that the 2020 targets would reduce GHG emissions associated with the transportation sector in California by 10 percent in 2020, or 23 million metric tons of carbon dioxide equivalent.[33]

Market Changes and Key Environmental Regulations Likely Affected the Refining Industry

Stakeholders we interviewed identified three major changes that have likely recently affected the domestic petroleum refining industry.[34] First, crude oil production in the United States and Canada has increased, which has lowered the cost of purchasing crude oil for some refiners but poses some challenges related to crude oil transportation infrastructure constraints and the types of crude oils produced. Second, after many years of generally increasing domestic consumption of petroleum products, consumption has fallen since 2005, resulting in a smaller domestic market for refiners. Third, two key environmental regulations— CAFE and GHG vehicle emission standards and the RFS—have likely recently contributed to declining consumption of petroleum fuels, and compliance with the RFS has increased costs for some refiners.

Increased Crude Oil Production Has Lowered Crude Oil Costs for Some Refiners

U.S. and Canadian crude oil production has increased in recent years, leading to lower crude oil costs for some refiners, according to several stakeholders we contacted. According to EIA data, U.S. production of crude oil reached its highest level in 1970 and generally declined through 2008, reaching a level of almost one-half of its peak. During this time, the United States increasingly relied on imported crude oil to meet growing domestic energy needs. However, recent improvements in technologies have allowed companies that develop petroleum resources to extract oil from shale formations that were previously considered to be inaccessible because traditional techniques did not yield sufficient amounts for economically viable production. In particular, the application of horizontal drilling techniques and hydraulic fracturing—a process that injects a combination of water, sand, and chemical additives under high pressure to create and maintain fractures in underground rock formations that allow oil and natural gas to flow—have increased U.S. crude oil and natural gas production.[35] As shown in figure 7, monthly domestic crude oil production has increased by over 55 percent through September 2013 compared with average production in 2008. According to EIA, increases in production in 2012 and 2013 were the largest annual increases since the beginning of U.S. commercial crude oil production in 1859. Much of the increase in crude oil production has been from shale and other formations, such as the Bakken in North Dakota and the Eagle Ford in Texas, according to EIA data. Similarly, crude oil production in Canada— the largest foreign supplier of crude oil to the United States—has also increased significantly in recent years. From 2005 through 2012, total Canadian crude oil production increased by 32 percent and U.S. imports from Canada increased almost 50 percent.

The rapid growth in U.S. and Canadian crude oil production has lowered the cost of crude oil for some domestic refiners that have the access and ability to process these crude oils. For example, West Texas Intermediate crude oil—a domestic crude oil used as a benchmark for pricing—was $17.60 per barrel less expensive in 2012 than Brent, an international benchmark crude oil from the European North Sea that was historically about the same price as West Texas Intermediate. Those refineries able to take advantage of these lower priced crude oils have benefited because crude oil costs are the largest cost for refiners.[36] However, all refineries may not have been able to take advantage of these crude oils to the same extent for two key reasons:

- *Transportation infrastructure challenges*. The development of domestic and Canadian crude oil production has created some challenges for U.S. crude oil transportation infrastructure because some of the growth in production has been in areas with limited transportation linkages to refining centers. Most of the system of crude oil pipelines in the United States was constructed in the 1950s, 1960s, and 1970s to accommodate the needs of the refining sector and demand centers at that time. According to DOE officials, this infrastructure was designed primarily to move crude oil from the South to the North, but emerging crude oil production centers in Western Canada, Texas, and North Dakota have strained the existing pipeline infrastructure. Though pipeline capacity has increased—investments increased pipeline capacity to deliver crude oil to a key Cushing, Oklahoma hub by about 815,000 barrels per day from 2010 through 2013—EIA reported that it has been inadequate. Because of these challenges, some refineries may not have been able to take full advantage of crude oil production increases or had to rely on other more expensive crude oil transportation options such as truck, rail, or barge. For example, two of the refineries we visited recently installed facilities to enable them to receive crude oil from North Dakota or Canada by rail. According to EIA data, while refinery receipts of crude oil by these methods of transportation is a small percentage of total receipts, they have increased 57 percent from 2011 to 2012. Infrastructure constraints have, according to EIA, contributed to discounted prices for some domestic crude oils.
- *Configuration constraints at refineries*. Increasingly, the crude oil being produced in the United States and Canada has different characteristics from the crude oils that some domestic refineries are configured to use. Production of new domestic crude oil has tended to be light and sweet, whereas a portion of new Canadian production has been heavy and sour crude oils.[37] To a certain extent, some refineries can use these crude oils, but some have invested in new equipment in order to do so. For example, representatives of one refiner told us they had invested over $2.2 billion in a project including a new coking unit at a refinery to refine heavier and more sour crude oil from Canada.

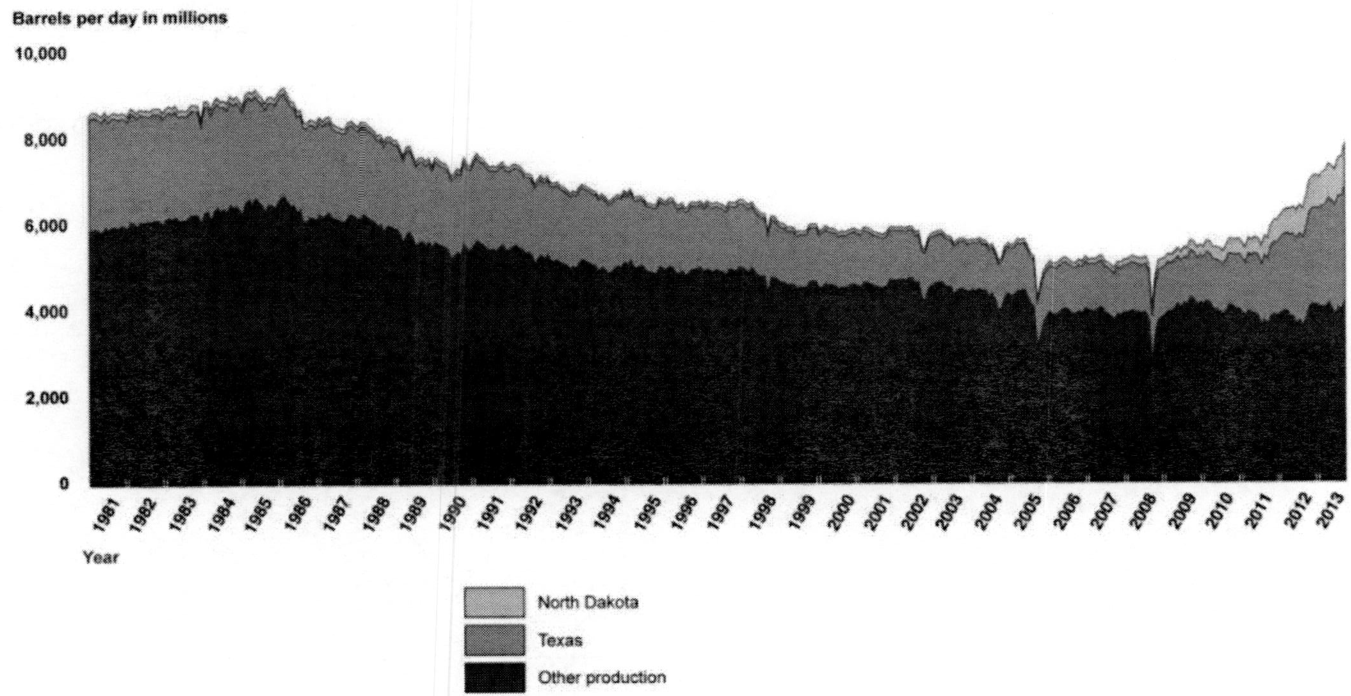

Source: GAO analysis of Energy Information Administration data.

Figure 7. Monthly Crude Oil Production in Texas, North Dakota and the Rest of the United States, January 1981-September 2013.

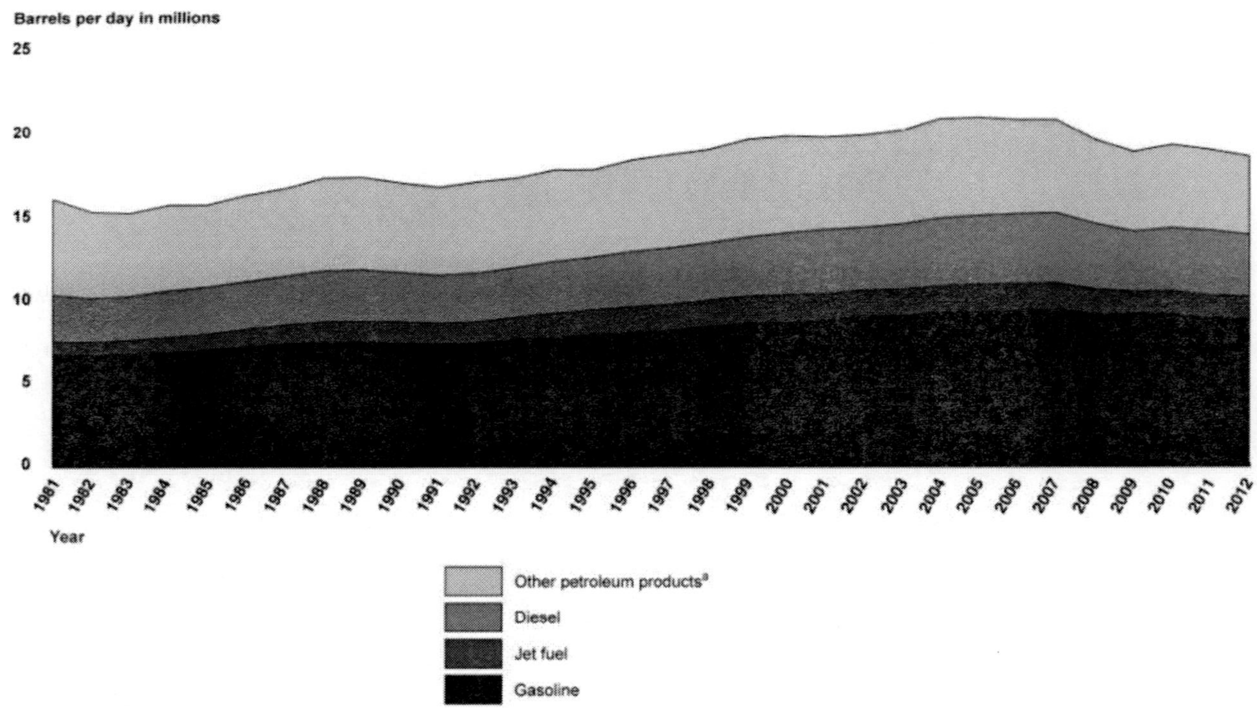

Source: GAO analysis of Energy Information Administration data.

Note: Data presented here include final consumption of fuels including petroleum-based fuels that have been blended with renewable fuels.

[a] Other petroleum products include kerosene, natural gas liquids, and other petroleum products.

Figure 8. Domestic Consumption of Gasoline, Diesel, and Other Petroleum Products, 1981-2012.

Domestic Consumption of Petroleum Products Has Declined

After decades of generally increasing domestic consumption of petroleum products, consumption has declined since 2005, resulting in a smaller domestic market for refiners, according to several stakeholders we contacted. Overall, consumption of gasoline, diesel, and other petroleum products in the United States increased from 1983 through 2005. In 2007, EIA projected that U.S. consumption would increase by nearly 30 percent between 2005 and 2030. As we reported in late 2007, trends in domestic refining capacity had not kept pace with consumption growth, though it was unclear whether and for how long that market tightness would continue.[38] However, as shown in figure 8, domestic consumption of petroleum products overall peaked in 2005 at 20.8 million barrels per day, and it declined by 11 percent through 2012. Consumption of gasoline, diesel, and jet fuel peaked in 2007 and declined by over 8 percent through 2012. More recent data indicate that these trends may now be starting to shift, as EIA estimated that petroleum product consumption increased in the first 11 months of 2013 compared with the first 11 months of 2012.

According to several stakeholders we contacted and information we reviewed, a number of factors can affect consumption of petroleum products, including economic activity and crude oil and petroleum prices. For example, the recession of 2007 to 2009 reduced economic activity and demand for gasoline, and historically high gasoline prices have discouraged the use of gasoline. Stakeholders and information we reviewed also cited the effect of more stringent fuel economy standards and the RFS, which are discussed in the next section.

Several stakeholders told us that this broad shift from growing to falling consumption of petroleum products has affected the domestic refining industry because it has resulted in a smaller domestic market. The U.S. market is important for domestic refineries because U.S. refiners have historically primarily sold their products domestically. On average, the United States exported almost 1 million barrels per day of domestic petroleum products from 2000 through 2005—less than 6 percent of U.S. refinery production. As discussed below, the refining industry has shifted sales to export markets amid a declining domestic market.

Two Key Regulations Have Likely Contributed to Declining Fuel Consumption and Compliance with One Has Increased Some Refiners' Costs

According to stakeholders and the information we reviewed, two recently strengthened key environmental regulations—the coordinated CAFE and GHG vehicle emission standards, and the RFS—have likely affected the refining industry by reducing the consumption of petroleum fuels, and compliance with the RFS has recently increased costs for some refiners, as well as other challenges. The other three key environmental regulations we reviewed have had minimal effects to date because they have either not yet been implemented or have generally not affected the industry in a major way, according to several stakeholders and information we reviewed.

CAFE and GHG Vehicle Emission Standards

According to information we reviewed and two stakeholders we contacted, CAFE and GHG vehicle emission standards have contributed to reductions in the consumption of petroleum fuels, but the extent is unclear. These standards aim to reduce oil consumption, and although they do not require changes at the refinery level, they can affect refineries indirectly by contributing to improvements in the overall efficiency of the vehicle fleet and, therefore, reducing fuels consumption. However, the National Academy of Sciences reported that it is difficult to isolate the effect of CAFE and GHG vehicle emissions standards from other factors that also affect consumption, such as higher fuel prices and changing driving habits.[39]

Stakeholders had different views on the extent to which CAFE and GHG vehicle emission standards have affected consumption of petroleum products. We reported, in 2007, that CAFE standards—along with higher fuel prices and other factors—contributed to a reduction in transportation fuel consumption of 2.8 million barrels per day in 2002.[40] CAFE standards for cars largely did not change from 1990 through 2010, but they were strengthened beginning with model year 2011. According to EPA and DOE officials, since the standards did not change until recently, CAFÉ and GHG vehicle emissions standards did not cause the shift from growing consumption to declining consumption discussed previously. Regarding the strengthened standards, EPA estimated in 2010 that vehicles were expected to save 1.3 billion gallons of gasoline in

2013 compared with model year 2011 standards.[41] This is equivalent to about 1 percent of EIA's projection of gasoline consumption in 2013. A stakeholder told us that the CAFE and GHG vehicle emissions standards have likely had a relatively large impact on petroleum demand declines in the past few years, but it is unclear how much declining demand overall can be attributed to these standards versus other factors such as the recent economic recession and higher fuel prices. On the other hand, EPA and DOE officials, and a refinery representative told us that the most recent changes to CAFE and GHG vehicle emissions standards have had a marginal effect on petroleum demand so far. DOE officials also told us that the impact of the standards has been limited because they affect new car sales, and there are a relatively small number of new vehicles in the overall fleet.

RFS

Several stakeholders we contacted and information we reviewed cited three main effects that the RFS has had on the domestic petroleum refining industry or individual refiners—compliance has increased costs, declining domestic consumption, and investment uncertainty. In addition, EPA has been late in issuing annual RFS standards, and several factors contribute to the delays.

RFS Has Had Three Main Effects

Stakeholders we contacted and information we reviewed identified three main ways the RFS has affected U.S. petroleum refiners: (1) compliance has recently increased costs for some refiners, (2) required blending of renewable fuels has contributed to declining domestic consumption of petroleum-based transportation fuels, and (3) EPA's delays in issuing annual RFS standards may have contributed to investment uncertainty for some refiners. First, compliance with the RFS has recently increased costs for some refiners, according to information we reviewed and several stakeholders we contacted. Under the RFS regulations, refiners and other obligated parties are required to ensure U.S. transportation fuels include certain amounts of renewable fuels. To comply, refiners generally have two options—they can purchase and blend renewable fuels themselves, or they can pay others to blend or use renewable fuels by purchasing credits.[42] These credits can be freely traded, and prices for credits are established based on the market and generally reflect the stringency of requirements and the costs of incorporating additional renewable fuels into the transportation fuel system to comply with the RFS—if costs increase, credits prices would tend to increase as well. According to EIA, corn-based ethanol credit prices were low—between $0.01 and $0.05 per gallon between 2006 and much of 2012—because it was generally economical to blend up to or above the level required by the RFS. However, in 2013, prices for these credits increased to over $1.40 per gallon in July before declining to about $0.20 per gallon as of mid- November 2013.

Several stakeholders told us this increase in credit prices was primarily due to RFS requirements exceeding the capability of the transportation fuel infrastructure to distribute and the fleet of vehicles to use renewable fuels, referred to as the "blend wall." A refiner we spoke with also attributed the decline in credit prices in the second half of 2013 to EPA's statements expressing its desire to address the blend wall. We have previously reported on the blend wall and other challenges to the increasing use of renewable fuels.[43]

While the RFS applies to all refiners in the same way, the effect of the rise in credit prices may depend on each refiner's situation. However, in comments to this report EPA stated that refiners experience the same compliance costs. As a result of higher costs, several stakeholders told us refiners could reduce production, produce more jet fuel, which is not subject to RFS requirements, or increase exports to nations where the RFS does not apply. (See app. III for more information about the blend wall, RFS credits, and views on how they have affected U.S. refiners.)

Second, the RFS has contributed to the declining domestic consumption of petroleum-based transportation fuels. Under the RFS regulation, refiners and other obligated parties are required to ensure U.S. transportation fuels include certain amounts of renewable fuels. As a result, refiners and other industry participants have blended increasing amounts of renewable fuels. For example, consumption of ethanol has increased almost 8-fold since 2000, from 1.7 billion gallons in 2000 to 12.9 billion gallons in 2012. According to EIA, increased ethanol use since 2007 displaced over 4 billion gallons of petroleum-based gasoline, or about 3 percent of gasoline consumption in 2012. As discussed previously, decreases in consumption affect refiners by decreasing the size of the domestic market. Since the RFS was established in light of concerns about the nation's dependence on imported crude oil, decreased consumption of petroleum products may further some of the objectives of the RFS.

Third, the RFS has contributed to investment uncertainty for refiners according to several stakeholders because EPA has not issued annual RFS standards on time since 2009.[44] Beginning in calendar year 2009 and through calendar year 2022, EPA is required to set annual blending percentages for total renewable fuels, advanced biofuels, cellulosic biofuels, and biomass-based diesel fuels by November 30 of the preceding calendar year.[45] However, as shown in figure 9, EPA has missed the statutory deadline to set annual percentages since 2009.[46] Most recently, EPA issued 2013 standards in August 2013—over 8 months late—and has not issued the 2014 standards. EPA proposed the 2014 standards on November 29, 2013, and EPA officials told us that they plan to finalize the standards in Spring 2014. The RFS compliance period—the time during which refiners and other parties incur obligations under RFS and can take steps to incorporate additional renewable fuels to create credits for compliance—is set by statute to be a full calendar year, and delays do not change this compliance period.[47] As a result, when the RFS standards are issued late, the industry has less time to plan and budget effectively. Several representatives of refiners told us that delays in issuing annual RFS standards increase uncertainty for refiners and renewable fuel producers, making it more difficult to make long-term planning decisions. One refining company representative told us that the company has reduced capital investments due to uncertainty related to the RFS. In contrast, EPA officials told us that there is no indication that delays have caused significant problems for refiners. They also noted that delays could actually make annual standards more robust since EPA then has more data upon which to base decisions.

Regulatory Development Processes Contribute to EPA Delays in Issuing RFS Standards

EPA officials told us that delays in issuing RFS standards have largely been due to the length of the regulatory development process, which includes interagency and public reviews. Under the interagency review process, EPA is to follow certain procedures before publishing proposed or final regulations that establish annual RFS standards, including submitting draft proposed and final regulations to the Office of Management and Budget (OMB), which coordinates review of the draft regulations by other agencies, as well as conducting its own review. The interagency review process is to ensure that regulations are consistent with the President's priorities, among other things, and that decisions made by one agency do not conflict with the policies or actions taken or planned by another. Under the public review process, EPA must publish a proposed standard in the Federal Register, provide the public with the opportunity to review and comment on the proposal, and address comments received before finalizing the regulation. According to EPA officials, the interagency and public review processes can be time consuming because the RFS standards involve complex and controversial issues and balance competing agricultural, energy, and environmental policy interests.

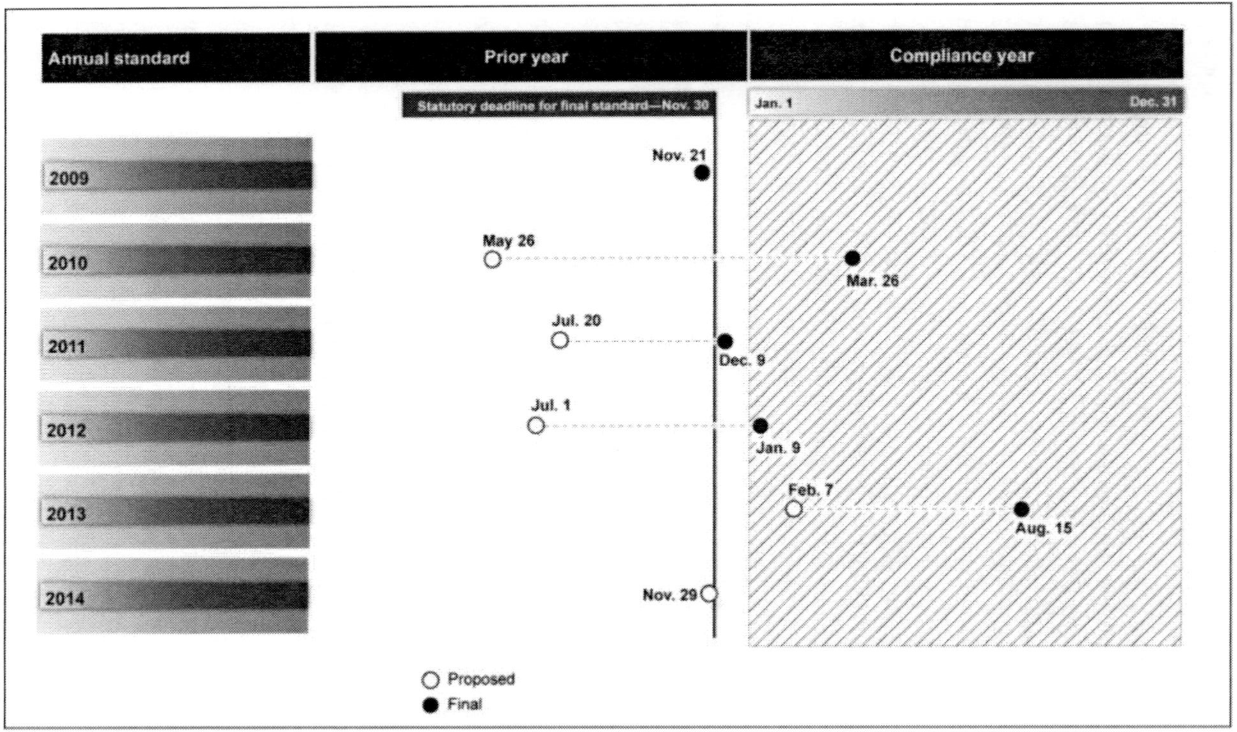

Source: GAO analysis of proposed and final regulations.

Note: The proposed and finalized dates are the regulations' publication dates in the Federal Register, not the dates EPA signed the regulations and publicly released them.

Figure 9. Timeline of EPA's Issuance of Proposed and Final RFS Standards (2009-2014).

Examples of the Effect of Increasing Renewable Fuel Standard (RFS) Credit Costs on Refiners

- According to company financial regulatory filings, the cost of credits increased almost 10-fold in the first 9 months of 2013 for one refiner compared with a similar period in 2012, and credit costs for another refiner more than doubled.
- A refining company representative told us that credit costs under the RFS exceeded the company's entire operating budget, including labor and electricity, for a period in 2013.
- As a result of higher credit costs, representatives of a refiner told us they have begun to consider the point at which their revenue may cease to cover the cost of purchasing credits. Representatives told us that, at a certain point, the company might need to decrease production levels, and thereby reduce its renewable fuel blending obligations.
- An ExxonMobil official said that RFS credit costs did not have a significant impact on the company's financial performance during the second quarter of 2013 because ExxonMobil meets the majority of its obligation by blending renewable fuels itself.

Source: GAO.

In 2009, we recommended that EPA and other agencies track their performance for developing significant regulations against targeted milestones to identify opportunities for improvement.[48] We found that monitoring actual versus estimated performance could help agency managers identify steps in the process that account for substantial time and provide information necessary to evaluate whether time was well spent. In this regard, EPA stated in comments to our 2009 report that it uses an agency-wide Action Development Process that tracks 14 milestones as it develops proposed rules and additional milestones in developing final regulations. For example, EPA tracks when its senior management approves of a document describing the scope of the regulation and the analytical work necessary to develop it, known as the detailed analytic blueprint. In comments to our 2009 report, EPA stated that it used an internal tracking system along with additional information to develop regulatory management reports to EPA managers and executives. EPA stated at the time that this process helps management identify regulations that are off-track so that corrective steps can be taken to expedite their completion. EPA officials told us that they develop RFS regulations using the same procedures used for developing all EPA regulations. However, even with these systems, EPA has not met its statutory deadlines under the RFS in the five annual standards since 2009. EPA has not conducted a systematic review of its experience issuing RFS regulations to identify the underlying causes of repeated delays and has not identified changes in its approach that may help to avoid these delays in the future. Without such analyses and a plan to address the underlying causes of the delays, EPA risks repeating them.

Other Key Regulations

The other key regulations that we reviewed—Tier 3 standards, stationary source GHG requirements, and LCFS—have had minimal effects to date because these regulations either have not yet been implemented (Tier 3 standards) or, with respect to the other two, have not affected industry operations or costs in a major way, according to stakeholders and information we reviewed. Specifically:

- *Tier 3 standards*. Tier 3 standards were proposed on May 21, 2013, and EPA announced final standards on March 3, 2014; therefore, they have had not had a direct effect on industry to date.
- *Stationary source GHG requirements*. Representatives of two refiners told us that stationary source GHG requirements have been burdensome to refiners; however, several other stakeholders told us they have not had a major effect, and EPA officials told us they were aware of only three refineries that have received major source GHG permits since the GHG permitting program was implemented in 2011.[49] A refining company representative expressed concerns to us about the lengthy permitting process to authorize GHG emission increases. However, stationary source GHG requirements do not apply unless an existing refining facility proposes a major modification or a new refinery is proposed for construction.[50] An EPA official said that, in most cases, best available control technologies selected to comply with GHG requirements for refining facilities involve energy efficiency improvement measures that could help refiners reduce fuel consumption and save money. Further, EPA

officials also explained that, in some cases, delays can occur when the refinery applicant has not provided EPA with the proper information to proceed with processing the permit.[51] The Clean Air Act requires that EPA approve or deny such permits within 12 months of receiving a complete application.

- *LCFS.* CARB, the entity responsible for implementing LCFS, said the regulation has had a modest effect to date—increasing fuel prices by about $0.01 per gallon. LCFS is the subject of several ongoing lawsuits,[52] which resulted in a 4-month delay in some regulatory activities and uncertainty about the status of the regulation.[53] According to a study conducted by a consultant on behalf of industry, the ongoing legal challenge to LCFS is creating uncertainty that discourages new investments by industry.[54] A refining industry trade association representative told us many refiners that previously invested in new components for their California facilities to process heavy crude oils may not be able to make an adequate return on investment since the LCFS disincentivizes the use of carbon intensive heavy crude oils. However, a CARB official noted that LCFS does not specifically prohibit any crude oil from being processed in California refineries, but rather it ensures that the LCFS' goal to reduce carbon intensity in transportation fuels is not affected by increased use of higher carbon intensity crude oils. Nevertheless, California refiners have thus far been able to comply with LCFS requirements by blending lower carbon intensive renewable fuels—such as Brazilian sugar-cane ethanol—or purchasing carbon credits as an alternative method of compliance.

INDUSTRY OUTLOOK DEPENDS ON A NUMBER OF FACTORS

Stakeholders we contacted and information we reviewed generally suggest that the outlook of the U.S. refining industry depends on a number of factors, in particular: (1) future domestic consumption of petroleum products; (2) the extent to which key environmental regulations raise costs for domestic refiners; and (3) the extent to which domestic refiners will be able to export and compete in international markets.

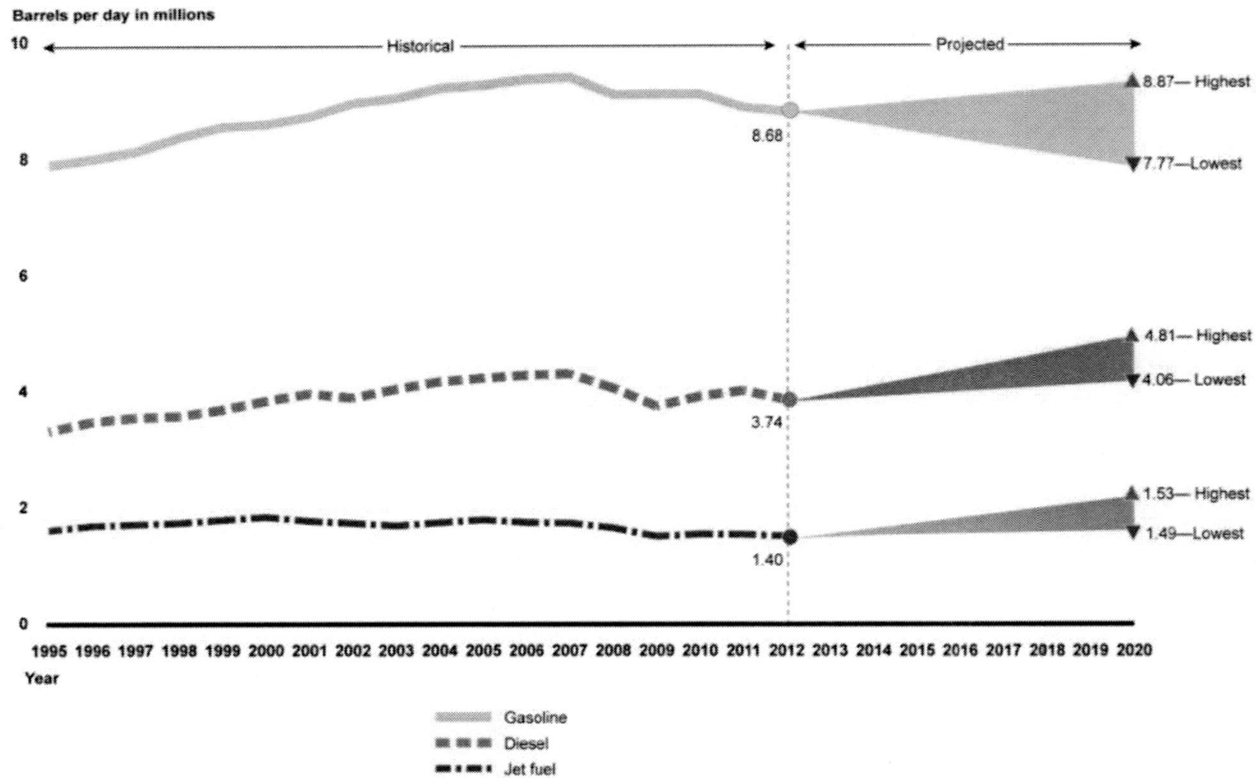

Source: GAO analysis of Energy Information Administration and IHS data.

Note: This figure displays the highest and lowest projections of U.S. consumption of gasoline, diesel, and jet fuel in the IHS and EIA forecast scenarios we reviewed. Specifically, for gasoline, the highest projection is IHS's forecast, and the lowest is EIA's "low or no net imports" scenario, which builds numerous assumptions to generate a hypothetical situation in which the United States eliminates net imports of liquid fuels. For diesel, the highest projection is EIA's "high economic growth" scenario, which assumes stronger growth in gross domestic product than in the reference scenario, a business- as-usual estimate that assumes current laws and policies are unchanged. The lowest projection for diesel is IHS. For jet fuel, the highest projection is EIA's "high economic growth" scenario, and the lowest is EIA's 2014 "reference" scenario.

Figure 10. Historical (1995-2012) and Projected (2020) U.S. Consumption of Gasoline, Diesel, and Jet Fuel.

Uncertain Future Domestic Consumption

The outlook of the U.S. refining industry depends on future domestic consumption of petroleum products, which is uncertain, according to stakeholders we contacted and information we reviewed. As discussed above, domestic petroleum product consumption declined by 11 percent from 2005 through 2012, and forecasts we reviewed project consumption of three major petroleum products—gasoline, diesel, and jet fuel—will be stable to slightly increasing through 2020, but not returning to high levels of the past.[55] Most of the scenarios in forecasts we reviewed from IHS and EIA project total consumption of gasoline, diesel, and jet fuel to increase slightly by 2020, with projections ranging from a decline of 2 percent to an increase of 7 percent compared with 2012 consumption.[56] Expectations differ somewhat by fuel, with all EIA scenarios projecting gasoline consumption to decline or remain stable and diesel and jet fuel consumption to increase from 2012 to 2020 (see Figure 10). IHS projects an increase in the consumption of both gasoline and diesel, with more robust growth projected for diesel. Scenarios in the forecasts we reviewed generally project consumption to decline after 2020.

Forecasts indicate that the level of future domestic consumption—the size of the domestic market for petroleum products—may affect future U.S. refinery production. In higher consumption scenarios, EIA's projections suggest higher refinery production than in scenarios with lower domestic consumption. Specifically, EIA projects that inputs to refineries—which track trends in refinery production—in 2020 would be about 1 million barrels per day higher in scenarios with higher domestic consumption—a difference of about 7 percent of 2012 inputs. This difference is equivalent to about eight average-size U.S. refineries.

Several stakeholders we contacted and information we reviewed highlighted various factors that can affect future domestic consumption levels and thereby the size of the largest market for the production of U.S. refineries, including the following:

- *Economic growth.* Faster economic growth tends to increase consumption, and EIA's forecast scenario with higher economic growth assumptions projects greater future consumption of petroleum products than a scenario with low economic growth.
- *Crude oil and petroleum product prices*. Higher prices for crude oil and petroleum products tend to decrease consumption. For example, of the forecast scenarios we reviewed, the scenario that assumes high

future oil prices projects lower domestic consumption of petroleum products.

- *Shifts in consumer behavior and demographic trends.* Changes in consumer behavior, such as reduced driving, along with demographic trends, such as an aging population and fewer young people with driving licenses, may reduce future consumption, according to EIA.
- *Key regulations.* Three of the key regulations we reviewed—CAFÉ and GHG vehicle emission standards, RFS, and LCFS—are expected to reduce domestic consumption of petroleum products in the future according to information we reviewed, though it is uncertain by how much. CAFE and GHG vehicle emission standards will require more efficient vehicles in the future, which may reduce future consumption of fuels. EPA estimated that the model year 2012-2025 standards are projected to reduce U.S. consumption of crude oil by 2.2 million barrels per day by 2025, equivalent to almost 15 percent of crude oil used by refineries in 2012.[57] Similarly, under the RFS statute, unless waived by EPA, renewable fuels blending is required to double by 2022, which EPA estimated would reduce gasoline and diesel demand by 13.6 billion gallons, equivalent to about 10 percent of consumption in 2012.[58] Furthermore, CARB projected that the LCFS would help decrease future gasoline consumption in California. However, the extent to which these regulations will reduce future consumption depends on actions by regulators and market and other developments. For example, as discussed above, EPA has proposed to reduce renewable fuel requirements for 2014 due to an inadequate supply in light of the blend wall and other issues. EPA stated that the framework it applied to determine the proposed percentage standards could be appropriate for later years. Therefore, the potential for RFS to reduce petroleum-based fuel consumption will depend on the percentages finalized by EPA, which, in turn will depend on the development of advanced renewable fuel sources and market infrastructure and could be affected by legal challenges, as well as any legislative action to modify the RFS.[59]

Costs of Key Regulations

Stakeholders we contacted and information we reviewed generally suggest that the outlook of the U.S. refining industry will also depend on the extent to which some key regulations—RFS, Tier 3, stationary source GHG requirements, and LCFS—increase costs for refiners. In general, increasing costs for refiners may be absorbed by refiners themselves (i.e., by reducing their profits), be passed on to consumers through higher product prices, or both. The requirements on domestic refiners from the key regulations we reviewed generally are expected to collectively have a greater effect in the future, for example, by affecting more refiners (such as the stationary source GHG requirements to the extent that more refineries make modifications over time), or becoming more stringent in the future (such as the RFS), potentially increasing costs for refiners. In addition, several stakeholders told us that the uncertainty surrounding these regulations—and what costs they will impose—can affect the market climate within which refiners and other market participants make investment decisions such as whether to expand a refinery's ability to process different crude oils, or to build new advanced biofuel processing facilities. Such uncertainty can discourage investments in the industry overall.

- *RFS.* RFS may increase costs for some refiners depending on the percentages of renewable fuels required by EPA and on other factors. As discussed previously, costs for some refiners to comply with RFS rose in 2013, which some of the stakeholders we contacted attributed to concerns about the blend wall. The blend wall may remain a concern into the future because statutory renewable fuel blending requirements continue to increase—they more than double from 2012 to 2022—while the consumption of petroleum products is expected to increase only slightly. Several stakeholders told us that the effect of the RFS depends in particular on how EPA addresses the blend wall in the annual standards it issues in the future. Furthermore, EPA's timeliness in issuing the standards could also affect costs to the extent that delays affect the supply of renewable fuels, RIN prices, and refiners' ability to plan and budget effectively for compliance. Several representatives of refiners told us that future delays would contribute to investment uncertainty and higher costs for refiners. EPA officials said that they did not believe delays have affected market participants, and that the market for RFS credits has provided flexibility to refiners and other obligated parties. The extent to which the RFS increases costs in the future could also be affected by the outcome of any relevant litigation and of legislative proposals to change the RFS or how EPA implements it.

- *Tier 3*. According to EPA, to meet the Tier 3 fuel sulfur standards, refiners would need to install or upgrade hydrotreating capacity or take other steps to reduce the sulfur content in fuels, which will likely increase industry-wide costs.[60] EPA projected that 67 out of 108 refineries would modify or purchase some equipment, and the capital costs of installing this equipment and operating costs to run it would average about $0.0065 per gallon, and total $804 million in 2017. An industry study of the Tier 3 proposal estimated that the regulation would increase costs by up to $0.09 per gallon for the highest-cost refinery, and several refinery representatives told us that Tier 3 would increase their costs.[61] The extent to which the refining industry will be affected by Tier 3 standards would have been greater had EPA decreased the per-gallon maximum allowable sulfur levels in gasoline—known as caps. In the final standards, EPA maintained the current 80 ppm cap, but had sought comment on whether it should decrease the cap to as low as 20 ppm. A stakeholder told us that Tier 3 would be manageable if EPA maintained the current caps, but far more difficult if the caps were lowered. One study we reviewed estimated that industry could incur additional capital expenses to achieve lower sulfur caps, ranging from $2 billion to over $6 billion dollars depending on the sulfur cap level in the final standards.[62]
- *Stationary source GHG requirements*. Several of the refining industry representatives we contacted expressed concerns that stationary source GHG requirements could become more stringent in the future. The current permitting framework is a case-by-case determination that takes into account costs, among other factors, and places no requirement on existing refineries unless they undertake a major modification. However, EPA entered a settlement in which the agency agreed to develop national performance standards—called New Source Performance Standards—for GHG emissions from new and modified refineries, and GHG emissions guidelines for certain facilities at existing refineries.[63] Although EPA has no current schedule to issue these standards, EPA committed in its settlement to issue them, and several stakeholders expressed concern that future standards could impose more strict controls involving higher costs at refineries. A stakeholder told us that until EPA clarifies its approach for NSPS, many refiners will be reluctant to make certain investments in their refineries out of concerns that their investments may be

unprofitable given future requirements. In addition, some companies may preemptively factor in the cost of emissions control technologies in their investment analyses.

- *LCFS*. Two stakeholders and a refining industry trade association told us that California refiners could face higher costs or compliance challenges unless CARB adjusts future low carbon fuel requirements. CARB has estimated that the cost of LCFS on gasoline and gasoline-substitute fuels is likely to range between an increase of $0.09 per gallon, and a decrease of $0.13 per gallon by 2020.[64] However, an industry study estimated that the LCFS could cost the refining industry an average of $0.70 per gallon by 2020.[65] The study also projected that 5 to 7 of 14 California refineries could cease production by 2020, and the LCFS could raise other compliance challenges because of insufficient supplies or consumer uptake of cellulosic, Brazilian sugar- cane ethanol, and other low carbon intensity fuels or vehicle technologies. CARB officials told us that if it proves more difficult than expected to meet LCFS requirements, CARB could introduce cost containment provisions to increase the availability of credits, such as through a "safety valve" to release additional credits at a set price, or by providing extra credits to certain compliance approaches.[66] Stakeholder told us that the decisions CARB makes with respect to the LCFS may affect California refiners' ability to stay in business and compete with refiners in other states and countries.

Extent to Which Refiners Can Export and Compete in Foreign Markets

While the domestic refining industry has increasingly relied on export markets, stakeholders and forecasts we reviewed indicate that the industry's future competitiveness is uncertain and that foreign markets present both challenges and opportunities for U.S. refiners.

Refiners Projected to Continue to Rely on Foreign Markets

Forecasts and data we reviewed from EIA and IEA suggest that future domestic refinery production levels may depend on exports of petroleum products.[67] Petroleum products are increasingly global commodities, and EIA data indicate that as domestic consumption has declined, refiners have looked to foreign markets to sell products. Since 1949, the United States had been a net importer of petroleum products, but this long-term trend reversed in 2011 when the United States became a net exporter of total petroleum products. According to EIA data, the United States recently exported more petroleum products than other leading exporters, including Russia, India, and Singapore, and petroleum product exports have represented an increasing share of U.S. refinery production. Exports of petroleum products represented 7 percent of refinery production in 2007 but increased to 17 percent in 2012. Major markets for U.S. exports include Central and South America, Mexico, and Europe, to which U.S. refiners sent nearly all diesel exports in 2012. The United States exports more diesel than gasoline, though U.S. refiners have been increasing exports of gasoline to Central and South America and Africa.[68]

Forecasts that we reviewed generally project that exports will remain strong. According to most of the EIA and IHS forecasts we reviewed, exports of petroleum products are expected to increase until 2015, but the extent of the increase is unclear. As shown in figure 11, EIA scenarios project export levels from 2.6 to 3.4 million barrels per day by 2020, a relatively wide range.[69] Even the lowest projection for petroleum product exports in 2020 is above 2010 levels, indicating a general expectation that exports will remain strong.

Foreign Markets Could Present Both Challenges and Opportunities for U.S. Refiners

The extent to which domestic refiners are able to export their products will depend on the competitiveness of domestic refiners compared with foreign refiners, and stakeholders we contacted and information we reviewed highlighted both challenges that may inhibit competitiveness and opportunities that may increase it in the future. To sell products abroad, refiners need to be competitive—that is, they must be able to supply fuels that foreign purchasers want to buy at prices that are attractive. Stakeholders and information we reviewed suggest that various factors may affect the U.S. refining industry's future competitiveness, including: (1) the balance between global refining capacity and global demand for petroleum products, (2) costs associated with environmental regulations, (3) exports to nations with stringent fuel standards, and (4) increasing domestic and Canadian crude oil production. More specifically:

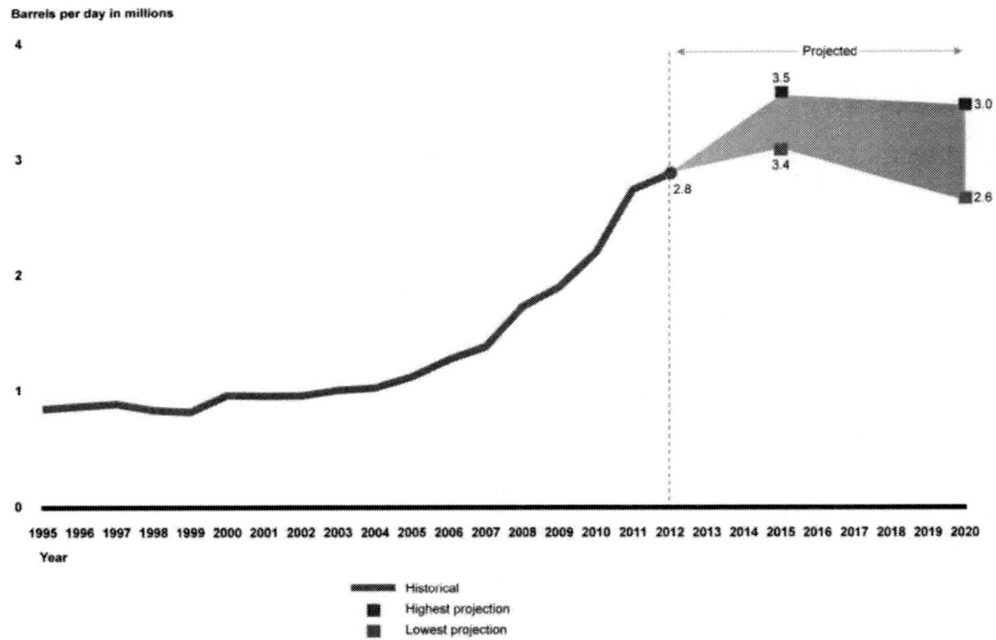

Source: GAO analysis of Energy Information Administration and IHS data.

Note: This figure displays the highest and lowest projections of U.S. petroleum product exports by the EIA forecast scenarios we reviewed. Specifically, the highest projection is from EIA's "greenhouse gas $25 and low gas prices" scenario, which assumes a fee for carbon emissions starting at $25 per metric ton in 2014, as well as high crude oil and natural gas resources consistent with EIA's "high oil and gas resource case." The lowest projection is from EIA's "low oil and gas resource" scenario, which assumes lower crude oil and natural gas resources than in the "reference" scenario.

Figure 11. Historical (1995-2012) and Projected (2015 and 2020) U.S. Exports of Petroleum Products.

- *Balance between global refining capacity and demand.* IEA data indicate that competition from foreign refiners may increase as global refining capacity is projected to exceed global consumption, creating an imbalance between global supply and demand that may affect U.S. refiners. According to IEA, global consumption of petroleum products was about 78.9 million barrels per day in 2012 and is projected to grow an additional 6 million barrels per day by 2020—with growth concentrated in Asia and the Middle East and consumption declining in Europe. But IEA projects refining capacity may grow even faster, resulting in excess capacity (refining capacity beyond that needed to meet consumption) of nearly 15.5 million barrels per day by 2020, in contrast to an estimated 4.8 million barrels per day of excess capacity in 2012.[70] Excess refining capacity is likely to result in greater competition in foreign markets overall, and some regions may present particular challenges. A large share of the new capacity is expected in China, India, and the Middle East, and representatives of two refiners indicated concern that capacity additions in some of those regions may present competition for U.S. refiners. Several other stakeholders were optimistic about domestic refiners' ability to compete in the future. According to IEA, capacity additions in China and India are intended to keep pace with growing consumption in those regions, but new Middle Eastern refineries are intended to be export facilities and may present increased competition to U.S. refiners.
- *Costs associated with environmental regulations.* As discussed above, the key environmental regulations we reviewed could collectively impose additional costs on the domestic refining industry. These costs could affect the industry's ability to compete internationally to the extent that foreign refiners do not face similar costs. In addition, regulatory uncertainty can affect refiners' competitiveness if it inhibits the industry from making investments that would otherwise lower costs. Not all of the key regulations we reviewed would be expected to affect the industry's competitiveness. In particular, CAFE and GHG vehicle standards do not impose requirements on refiners. In addition, although RFS and Tier 3 standards could impose requirements with potential associated costs on some refiners, they would not apply to exported fuels—they apply only to fuels sold in the United States regardless of where they are produced.

- *Potential for increased exports to nations with stringent fuel standards*. In general, U.S. refineries are among the most sophisticated in the world and have generally been optimized to produce large proportions of cleaner-burning gasoline. IEA has pointed out that refiners in many parts of the world face challenges producing fuels that meet high product quality and environmental performance standards.[71] Therefore, some U.S. refiners may benefit from any trend toward higher quality and more stringent environmental performance standards. In this regard, actions refiners may take to reduce gasoline sulfur to comply with proposed Tier 3 standards could enable them to export to markets—such as Japan and much of Europe—that already require low sulfur gasoline.[72] On the other hand, representatives of a refiner pointed out that refiners could undertake such investments on their own—without Tier 3—if such exports were sufficiently economically attractive.
- *Increasing domestic and Canadian crude oil production*. As discussed above, increasing U.S. and Canadian crude oil production has led to lower cost crude oil for some refiners, providing a competitive advantage.[73] All of the forecast scenarios we reviewed from EIA, IEA, and IHS anticipate increases in U.S. crude oil production, but the projections are uncertain and vary widely—from 6.8 to 9.8 million barrels per day in 2020 as shown in figure 12.[74] Projections are revised each year, and expectations for U.S. crude oil production in 2020 have increased in more recent forecasts. For example, the reference scenario in EIA's most recent forecast projects domestic crude oil production to approach a historical high of 9.6 million barrels per day in 2020, higher than the reference scenario from the prior year.[75] Canadian crude oil production—which accounted for about 16 percent of the crude oil used by U.S. refineries in 2012—is expected to increase as well: the reference scenario in EIA's international forecast projects that Canadian petroleum liquids production will increase more than 30 percent from 2012 levels, reaching approximately 5 million barrels per day in 2020.[76] The extent of the increase in future crude oil production can have implications for future petroleum product exports. For example, EIA's scenario that assumes more domestic crude oil and natural gas resources projects higher export levels than a scenario that assumes low crude oil and natural gas resources. Several stakeholders told us that various issues could mitigate U.S. refiners' ability to take advantage of growing

crude oil supplies. In particular, it is unclear whether planned expansions in pipelines and rail transportation will keep pace with growing production, and these infrastructure expansions could be affected by regulatory actions to address pipeline and rail safety.[77] Similarly, several stakeholders told us that potential future increases in crude oil exports, which are currently minimal, could put pressure on regional crude oil prices, reducing the price advantage of U.S. refiners.[78]

CONCLUSION

The domestic petroleum refining industry has been and is expected to continue to be affected by several profound changes. Some of these changes, such as the growth in crude oil production in the United States and Canada, are reshaping the industry and creating new business opportunities. To take advantage of some of these opportunities, refiners and other market participants will need to invest—to upgrade refineries to be able to process different crude oils or to build pipelines or rail connections to move more crude oil from production to refining centers. Uncertainty can affect the market climate within which these investment decisions will be made. In this context, EPA's timeliness in issuing annual percentage standards under the RFS is important to help inform the investment decisions of the refining industry. In issuing annual percentage standards, EPA may waive the statutory volumes in whole or in part according to statutory criteria, which EPA has identified as potentially factoring in the blend wall, market developments, and other issues. However, EPA has missed the annual deadline for issuing annual standards under the RFS in most years. EPA has some systems in place to monitor and evaluate progress in developing regulations, which could provide useful information for understanding delays in RFS. But EPA has not identified the underlying causes of delays, and it has not developed a plan to address delays and, therefore, risks repeating delays. EPA delays in issuing RFS standards are important because delays do not change refiners' compliance periods accordingly and they therefore create uncertainty in the marketplace, potentially harming investment. Uncertainty among refiners, renewable fuel producers, and other market participants about how EPA will address the blend wall, which can be exacerbated by the prospect of litigation, can affect investment decisions and ultimately the availability and prices of the fuels they produce.

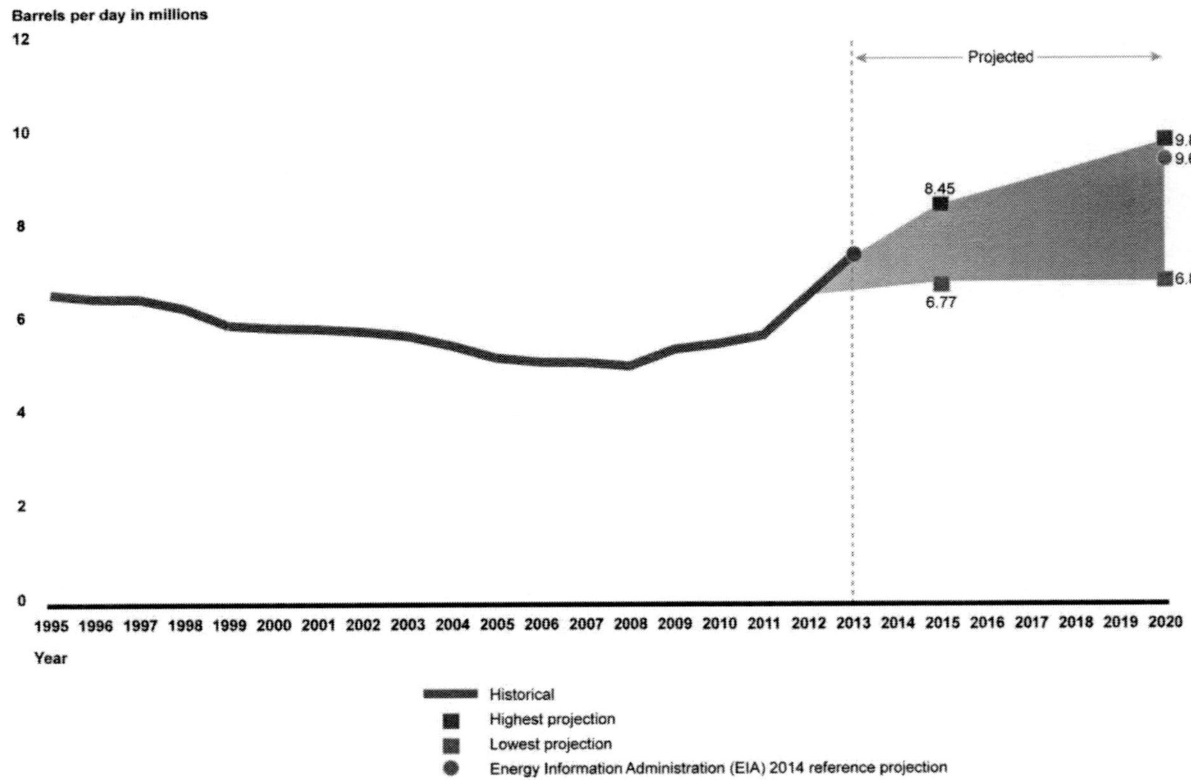

Source: GAO analysis of EIA and IHS data.

Note: This figure displays the highest and lowest projections of U.S. crude oil production by the IHS and EIA forecast scenarios we reviewed. Historical data for 2013 is the 9-month average of production through September. Specifically, the highest projection is

EIA's "greenhouse gas $10 and low gas prices" scenario, which assumes 100 percent higher oil recovery and 50 percent higher undiscovered resources than in the reference scenario, as well as a policy change that applies a $10 per metric ton fee for carbon emissions throughout the economy beginning in 2014. The lowest projection, EIA's "high net imports" scenario, assumes 50 percent lower crude oil recovery than the reference scenario, along with changes that would create the need for increased imports. Also displayed is EIA's 2014 forecast reference scenario that assumes current laws and regulations are unchanged.

Figure 12. Historical (1995-2013) and Projected (2015 and 2020) U.S. Crude Oil Production.

Recommendations for Executive Action

To improve EPA's ability to meet the annual statutory deadline for issuing annual RFS standards, we recommend that the Administrator of the EPA take the following two actions:

- Assess past experience to identify the underlying causes of delays in issuing annual RFS standards.
- Develop and implement a plan to address the causes of delays and help ensure RFS annual standards are issued on time.

Agency Comments and Our Evaluation

We provided drafts of this report to DOE, DOT, and EPA for review and comment. The three agencies provided technical comments on early or final drafts, which we incorporated as appropriate.

EPA also provided a letter in which it generally agreed with our findings and recommendations and clarified three topics discussed in the report.

First, regarding the effects of compliance with RFS, EPA asserted that refiners experience the same compliance costs regardless of whether they are fully integrated, with blending capabilities, or merchant refiners that purchase credits for compliance. Based on our work, we found the views of several stakeholders differed from EPA's. For example, in a 2011 study, DOE identified the degree to which a small refiner can actively blend production with renewable fuels is a large component that could contribute to economic hardship from compliance with the RFS.[79] In theory, market-based compliance systems—such as the RFS credit system—provide incentives for market participants to make decisions that would tend to equalize additional compliance costs over time. However, there can be physical infrastructure or contractual constraints, among various other factors, that could result in different outcomes in the short run. We added additional language to explain EPA's views in the report and in Appendix III.

Second, regarding the time-frame for RFS compliance, EPA stated that the RFS compliance deadline—the date by which refiners and other obligated parties must demonstrate compliance to EPA—is established through implementing regulations, not statute. EPA stated that it adjusted the 2013 deadline to provide additional time to demonstrate compliance. We acknowledge that EPA can extend the compliance deadline. However, the compliance period refers to the time during which refiners and other parties incur obligations under RFS and can take steps to incorporate additional renewable fuels to generate credits for compliance. This period is set by statute to be a full calendar year. We clarified language in the report to acknowledge EPA's ability to adjust the compliance deadline, essentially providing additional time for obligated parties to purchase credits, and its inability to adjust the compliance period.

Third, regarding Tier 3 standards, EPA announced the final standards while our draft was with the agency for comment. EPA stated that the final Tier 3 program is very similar to what it proposed, though EPA made some changes based on public input and updated its analyses. EPA provided technical comments to incorporate information from the final rule which we incorporated into the report, as appropriate. However, we were not able to obtain stakeholder and other views on the final Tier 3 rule for this report.

Frank Rusco
Director, Natural Resources and Environment

APPENDIX I: SCOPE AND METHODOLOGY

This report provides information on the domestic petroleum refining industry and its market and regulatory environment. Specifically, it addresses what is known about (1) major changes—including key environmental regulations—that have recently affected the domestic petroleum refining industry and (2) major factors that may affect the future of the domestic petroleum refining industry— including its production, profitability, and competitiveness in foreign markets.

To provide information on major changes that have affected the domestic petroleum refining industry and the future of the industry, we reviewed information including the following:

- studies by federal agencies and consultants,
- company financial regulatory filings, and
- proposed and final regulations and regulatory impact analyses.

To identify studies and other literature, we conducted searches of various databases, such as ProQuest and PolicyFile, for studies since published 2009.

We also asked agency officials and other stakeholders we contacted to recommend studies.

Based on our research and information from stakeholders, we identified five key regulations that were recently strengthened or proposed: (1) the Environmental Protection Agency's (EPA) Renewable Fuels Standard regulations, (2) the Department of Transportation's Corporate Average Fuel Economy and EPA's greenhouse gas vehicle emission standards; (3) EPA's Tier 3 Motor Vehicle Emission and Fuel Standards; (4) EPA's stationary source greenhouse gas requirements; and (5) the state of California's Low Carbon Fuel Standard. We reviewed agency regulatory impact assessments and industry and other studies on the effect of these regulations on industry. Other regulations may also affect the industry.

We also summarized the results of semistructured interviews with a nonprobability sample of 32 stakeholders. (See app. II for a list of these stakeholders.) Stakeholders included representatives from refining companies, environmental organizations, consultants, and officials from federal and state agencies. We also visited several refineries of selected refining companies. We selected these stakeholders to represent broad and differing perspectives on these issues based on recommendations from agencies and industry associations, along with other information. For example, to select refiners, we considered, among other factors, the size and location of their refineries, and whether they were vertically integrated or merchant refiners. When possible, we used a standard set of questions in interviewing stakeholders, including questions about the effect of the key regulations we reviewed. However, as needed, we also sought perspectives on additional questions tailored to these stakeholders' expertise and sought opinions from stakeholders on key issues, such as their views on the potential effects of exports on industry. Because we used a nonprobability sample, the views of these stakeholders are not generalizable to all potential stakeholders, but they provide illustrative examples of the range of views. Similarly, the conditions at the refineries we visited are not generalizable to all refineries. The stakeholder views we summarize were not necessarily supported by all types of stakeholders, though we identify differing views where appropriate. Stakeholders and information we reviewed identified a number of changes that have affected the industry and a number of factors that may affect its future, and we report on those that were most often cited.

To illustrate major changes over time and to describe the domestic petroleum refining industry, we summarized historical data from the Energy Information Administration (EIA) regarding such issues as capacity and location of refineries, crude oil production, and consumption of petroleum products. To assess the reliability of EIA data, we took several steps including reviewing documentation, interviewing EIA staff, and consulting with stakeholders. We determined the EIA data to be sufficiently reliable for the purposes of this report.

To provide information about the future of the domestic petroleum refining industry and major factors that could affect it, we also reviewed forecasts from EIA, the International Energy Agency (IEA), and IHS, and summarized projections through 2020 under different scenarios.[80] We selected these forecasts because they made projections through 2020, contained information broadly relevant to our report, covered multiple scenarios or offered a counterpoint scenario, and contained well- documented discussions of methodologies used and assumptions made.

While forecasts are subject to inherent uncertainties, we found these forecasts to be reasonable for describing a range of views about potential conditions of the domestic refining industry and major factors that will help determine these conditions. We reviewed and compiled data from relevant scenarios and compared them where appropriate. Specifically, we reviewed all 27 scenarios in EIA's 2013 forecast, the reference scenario in EIA's 2014 initial forecast,[81] and IHS's forecast, and, in particular, highlight the scenarios representing the highest and lowest projection of gasoline, diesel, and jet fuel consumption; petroleum product exports; and crude oil production.[82] We identified some differences in the metrics reported in the four forecasts and did not make direct comparisons in these instances.

We conducted this performance audit from November 2012 to March 2014 in accordance with generally accepted government auditing standards. Those standards require that we plan and perform the audit to obtain sufficient, appropriate evidence to provide a reasonable basis for our findings and conclusions based on our audit objectives. We believe that the evidence obtained provides a reasonable basis for our findings and conclusions based on our audit objectives.

APPENDIX II: LIST OF STAKEHOLDERS

Federal Agencies	1. Department of Energy[83]
	2. Department of Transportation
	3. Environmental Protection Agency
Refiners	4. Chevron

	5. CVR Refining
	6. HollyFrontier Corporation
	7. Marathon Petroleum Corporation
	8. Monroe Energy
	9. PBF Energy
	10. Philadelphia Energy Solutions
	11. Phillips 66
	12. Shell
	13. Tesoro Corporation
	14. Valero Energy Corporation
Other	15. American Fuel and Petrochemical Manufacturers
	16. American Petroleum Institute
	17. Baker & O'Brien Incorporated
	18. Prof. Severin Borenstein, University of California Energy Institute
	19. California Air Resource Board
	20. California Energy Commission
	21. Citigroup Global Markets Inc.
	22. Energy Policy Research Foundation (EPRINC)
	23. Environmental Integrity Project
	24. IHS
	25. Prof. Christopher Knittel, Massachusetts Institute of Technology
	26. Prof. Erich Muehlegger, Harvard Kennedy School
	27. Muse Stancil
	28. National Association of Clean Air Agencies
	29. Natural Resources Defense Council
	30. Sierra Club
	31. Stillwater Associates
	32. Turner, Mason & Company

Appendix III: Further Information Regarding the Renewable Fuel Standard, Compliance Credits, and the Blend Wall

To demonstrate compliance with the Environmental Protection Agency's (EPA) annual blending requirements under the Renewable Fuel Standard

(RFS), refiners use renewable identification numbers (RIN), which we refer to in this report as credits. A RIN is a unique 38-character code that renewable fuel producers and importers assign to each gallon of renewable fuel produced or imported. To demonstrate compliance with the RFS, refiners and importers must provide sufficient RINs for the volume of gasoline and diesel they produce for use in the contiguous United States and Hawaii. For example, to comply with the 2013 total renewable fuels standard requiring that renewable fuels compose at least 9.74 percent of gasoline and diesel, a refiner selling 100 million gallons of gasoline would have to provide 9.74 million total RINs to EPA. Refiners can obtain RINs by purchasing and blending renewable fuels themselves, or they can purchase RINs from renewable fuel producers, importers, blenders, other refiners, or other RIN-holding entities. RINs are valid for the calendar year in which they were generated, and up to 20 percent of a year's standard can be met with RINs from the previous year. Refiners and other obligated parties with more RINs than needed to meet the year's blending standard can hold them for use in the following year or sell them to another party that needs additional RINs to comply with the blending standard.

Prices for RINs reflect the cost of renewable fuels compared with the petroleum fuels they displace, the stringency of annual blending percentage standards, and other factors, and have varied over time. According to the Energy Information Administration (EIA), between 2006 and much of 2012, corn-based ethanol RIN prices were low—between $0.01 and $0.05 per gallon—because it was generally economical to blend up to or above the level required by the RFS. However, RIN prices for corn-based ethanol increased to over $1.40 per gallon in July 2013. Several stakeholders told us this increase in RIN prices was primarily due to RFS requirements exceeding the capability of the transportation fuel infrastructure to distribute and the fleet of vehicles to use renewable fuels, referred to as the "blend wall." EPA officials told us that high corn prices, which made ethanol more expensive relative to gasoline, also contributed to higher RIN prices during this period. These RIN prices have since come down to about $0.20 per gallon as of mid-November 2013. A refiner attributed this decline to EPA's statements expressing its desire to address the blend wall.

The blend wall exists because blending more than 10 percent ethanol with gasoline (called E10) is affected by constraints such as the limited availability of vehicles that can use higher ethanol blends.[84] In addition, higher ethanol blends are less widely available than E10 and must be priced at a discount to encourage greater consumption, according to EIA. EPA officials recently said the blend wall would be reached in 2014 when about 13.2 billion gallons of

E10 could be consumed.[85] Blending additional renewable fuels can be difficult and costly once the blend wall is reached because significant volumes of non-ethanol renewable fuels must be available, consumers must be encouraged to purchase additional higher blends of ethanol, and other market participants must develop the infrastructure to deliver those fuels.

Compliance with the RFS has recently increased costs for some refiners, according to information we reviewed and several stakeholders we contacted. While the RFS applies to all refiners in the same way, effects of rising or falling RIN prices may vary depending on each refiner's situation. For example, those refiners that have incorporated renewable fuel blending into their operations may have benefited from the rising prices relative to those refiners that are less well positioned. According to several stakeholders, RFS compliance has been most difficult for refiners with less of a retail presence, known as merchant refiners, because they do not blend their own fuel and must purchase RINs from others, increasing their cost of compliance. On the other hand, some industry participants may be relatively advantaged when the price of RINs rises. For example, an ExxonMobil official said that RIN costs did not have a significant impact on the company's financial performance during the second quarter of 2013 because ExxonMobil meets the majority of its obligation by blending its renewable fuels itself.

EPA officials told us, however, that the RFS program affects all refiners equally because obligations are the same regardless of whether refiners blend renewable fuels themselves or purchase RINs. In particular, EPA stated that refiners experience the same costs. If a company generates its own RINs, there is a cost associated with doing so, namely the cost for the renewable fuel compared to the petroleum fuel it displaces.

End Notes

1 See GAO, *Energy Markets: Increasing Globalization of Petroleum Products Markets, Tightening Refining Demand and Supply Balance, and Other Trends Have Implications for U.S. Energy Supply, Prices, and Price Volatility,* GAO-08-14 (Washington, D.C.: Dec. 20, 2007).

2 EIA is a statistical agency within the Department of Energy (DOE) that collects, analyzes, and disseminates independent information on energy issues.

3 We use the term key environmental regulations to refer to these five regulations.

4 The RFS was created in the Energy Policy Act of 2005, and then expanded through the Energy Independence and Security Act of 2007. See Pub. L. No. 109-58, § 1501 (2005) and Pub. L. No. 110-140, §§ 201-203, 210 (2007), codified as amended at 42 U.S.C. § 7545(o) (2013).

[5] IEA is an international organization composed of 28 of the member nations of the Organisation for Economic Co-operation and Development that, among other things, collects energy data and provides research and analysis on ways to ensure reliable, affordable, and clean energy.

[6] IHS is a firm that provides comprehensive economic and financial information on countries, regions, and industries.

[7] Forecasts reflect assumptions and data available at the time the forecasts are developed. The forecasts we reviewed were developed at different points in time. For example, the 2013 EIA forecast we reviewed was released in April 2013 while IHS's forecast was released in October 2013. More recent forecasts may more fully reflect recent developments such as changes in economic growth and in crude oil production.

[8] A refinery's capacity refers to the maximum amount of crude oil designed to flow into the distillation unit of a refinery, also known as the crude unit.

[9] These regions are Petroleum Administration for Defense Districts, which were created during World War II to help organize the allocation of petroleum products. Many petroleum data collection organizations use these regions to organize their data for analytical purposes.

[10] Coking processes use temperatures greater than 900 degrees Fahrenheit to thermally break molecules that make up feedstocks into products such as diesel, leaving behind petroleum coke—a coal-like material. Petroleum coke is used as a fuel input to produce electricity or as a raw material in the steel and aluminum industry.

[11] Except where noted, we use the term diesel to refer to what EIA calls distillate fuel oil. Distillate fuel oil is a general term that includes primarily diesel fuel, as well as fuel oils used for heating and electric power generation. In 2011, 90 percent of distillate fuel oil was diesel according to EIA data.

[12] GAO-08-14.

[13] Prices in this section were converted to 2012 dollars using the Bureau of Labor Statistics' Consumer Price Index.

[14] We have previously reported on several factors affecting these dynamics. See, for example: GAO, *Energy Markets: Estimates of the Effects of Mergers and Market Concentration on Wholesale Gasoline Prices,* GAO-09-659 (Washington, D.C.: June 12, 2009); GAO-08-14; and GAO, *Motor Fuels: Understanding the Factors That Influence the Retail Price of Gasoline,* GAO-05-525SP (Washington, D.C.: May 2, 2005). A more complete list of related reports is included at the end of this report.

[15] EPA may also reduce the requirements for total renewable fuel and advanced biofuel in years when EPA reduces cellulosic biofuel or biomass-based diesel under applicable provisions. See 42 U.S.C. § 7545(o)(7)(D)-(E) (2013). EPA also must provide public notice and an opportunity for comment before making a determination for a waiver.

[16] See GAO, *Biofuels: Potential Effects and Challenges of Required Increases in Production and Use,* GAO-09-446 (Washington, D.C.: Aug. 25, 2009) and GAO, *Energy- Water Nexus: Many Uncertainties Remain about National and Regional Effects of Increased Biofuel Production on Water Resources,* GAO-10-116 (Washington, D.C.: Nov. 30, 2009).

[17] Beginning in 2015, the RFS' statutory provisions call for more than half of advanced biofuels to be produced from cellulosic materials, which can include perennial grasses, crop residue, and the branches and leaves of trees. Such cellulosic biofuels must generally achieve at least a 60 percent reduction in life cycle greenhouse gas emissions compared with the gasoline or diesel fuel they displace. However, according to EIA, all EIA forecasts and projections made since the passage of the Energy Independence and Security Act of 2007 anticipated large shortfalls between the RFS targets and the volumes of cellulosic biofuels. In addition, some advanced biofuels must be produced from biomass-based diesel, which generally includes diesel made from biomass feedstocks such as soybeans.

[18] EPA's estimates of the benefits and costs of the regulations presented throughout this report refer to monetized amounts. As not all benefits and costs can be monetized, these may represent a subset of overall benefits and costs of the regulations. We did not independently assess EPA's estimates of the benefits or costs of these regulations. We are examining

EPA's economic analyses in relation to its adherence to guidance for regulatory analysis and expect to report later this year on these issues.

[19] The Energy Policy and Conservation Act, Pub. L. No. 94-163 tit. III (1975) established CAFE standards effective beginning in model year 1978.

[20] The NHTSA, under DOT, is responsible for setting and enforcing CAFE standards, among other things.

[21] The Energy Independence and Security Act of 2007, Pub. L. No. 110-140, tit. 1, subtit. A. Among other things, the act extended and codified NHTSA's reform of the CAFE standards by moving from a single fleet standard to an attribute-based standard for light trucks and passenger vehicles. The move from a single standard for a fleet to attribute- based standards for each vehicle model based on a vehicle's footprint was designed to address a number of downsides associated with moving to smaller vehicles, including potential safety implications and consumer choice limitations, see GAO, *Vehicle Fuel Economy: NHTSA and EPA's Partnership for Setting Fuel Economy and Greenhouse Gas Emissions Standards Improved Analysis and Should Be Maintained,* GAO-10-336 (Washington, D.C.: Feb. 25, 2010).

[22] In the coordinated rulemaking, NHTSA issued regulations establishing CAFE standards for model years 2012 to 2021 and identified nonbinding, projected standards for model years 2022-2025 due to a statutory requirement that allows the agency to set CAFE standards for not more than 5 model years in a given rulemaking. See 77 Fed. Reg. 62,624, 62,639 (Oct. 15, 2012). For further information on these rulemakings and the process used to set them, see GAO-10-336.

[23] The national program of CAFE and GHG vehicle emission standards is projected to result in an industry-wide fuel economy of 54.5 miles per gallon in 2025 if all GHG reductions are achieved with fuel economy technologies, but use of air conditioning refrigerant credits would lower the expected fuel economy to about 50 miles per gallon. Actual fuel economy will differ, depending on the fleet of vehicles sold and how manufacturers choose to meet the standards given certain flexibilities, including credits for improved air conditioning efficiency.

[24] In model year 2011, CAFE standards alone were in effect; from 2012-2025, both CAFÉ and GHG vehicle emission standards are in effect.

[25] Clean Air Act Amendments of 1970 §§ 202, 211(c)(1), Pub. L. No. 91-604, 84 Stat. 1676 (1970), codified as amended at 42 U.S.C. §§ 7521, 7545(c)(1) (2013) (commonly referred to as the Clean Air Act).

[26] For the proposed standards, see: 78 Fed. Reg. 29,816 (May 21, 2013). The final standards were signed by the Administrator of the EPA on March 3, 2014. (See: EPA, *Control of Air Pollution from Motor Vehicles: Tier 3 Motor Vehicle Emission and Fuel Standards, Final rule*, accessed March 5, 2014, http://www.epa.gov/otaq/documents/tier3/tier-3-fr-preamble-regs-3-3-14.pdf.) The Tier 3 standards replace the current vehicle emission and fuel sulfur standards that were established in 2000 under Tier 2. See 65 Fed. Reg. 6698 (Feb. 10, 2000). Tier 2 standards generally reduced sulfur in gasoline from 300 ppm to 30 ppm on an annual average basis and were phased in from 2004 through 2007.

[27] EPA also will provide a 3-year delay for certain small refiners, as well as certain small volume refineries that processed 75,000 or fewer barrels of crude oil per day averaged over 2011.

[28] EPA, *Control of Air Pollution from Motor Vehicles: Tier 3 Motor Vehicle Emission and Fuel Standards Final Rule; Regulatory Impact Analysis.* (EPA-420-R-14-005, March 2014).

[29] Several industry groups have challenged EPA's determination that its regulation of GHG emissions from new motor vehicles triggered permitting requirements under the Clean Air Act for stationary sources that emit GHGs. The U.S. Court of Appeals for the D.C. Circuit upheld EPA's determination, and the Supreme Court granted a petition to review the D.C. Circuit's judgment and will consider the issue in 2014.

[30] Permitting authorities include EPA regions and approved state and local agencies. Some states are not approved to issue GHG permits. In these states, EPA regions issue GHG permits.

[31] Permits that address GHG emissions from new or modified stationary sources are known as PSD permits. PSD permits address GHG emissions only for those sources and projects meeting specific emission thresholds. In addition, operating permits, also known as Title V permits, can be triggered by sources that emit GHGs at specified emissions thresholds. Operating permits generally do not add new pollution control requirements but contain procedural requirements such as compliance monitoring and reporting. Thus, only PSD permits will impose GHG emission control requirements.

[32] Specifically, the California Fifth District Court of Appeal allowed the LCFS regulations to remain in effect, but found that the manner in which the CARB adopted the LCFS violated certain procedural requirements of the California Environmental Quality Act and California's Administrative Procedures Act. As a result, CARB anticipates addressing the court's concerns with a regulatory package in 2014.

[33] CARB, *Staff Report: Initial Statement of Reasons Proposed Regulation to Implement the Low Carbon Fuel Standard, Volume I* (Sacramento, CA: Mar. 5, 2009).

[34] The stakeholder views summarized throughout this report were not necessarily supported by all types of stakeholders, though we identify differing views where appropriate.

[35] For more information, see: GAO, *Oil and Gas: Information on Shale Resources, Development, and Environmental and Public Health Risks,* GAO-12-732 (Washington, D.C.: Sept. 5, 2012).

[36] Closely related to increases in crude oil production, natural gas production has also increased in the United States in recent years. This increasing availability and lower prices for natural gas have been a benefit to domestic petroleum refiners since natural gas is an important energy source for many petroleum refineries, according to several stakeholders we contacted.

[37] Crude oil is generally classified according to two parameters: density and sulfur content. Less dense crude oils are known as "light," while denser crude oils are known as "heavy." Crude oils with relatively low sulfur content are known as "sweet," while crude oils with higher sulfur content are known as "sour." In general, heavier and more sour crude oils require more complex and expensive refineries to process the crude oil into usable products.

[38] See GAO-08-14.

[39] See: National Research Council, *Effectiveness and Impact of Corporate Average Fuel Economy (CAFE) Standards* (Washington, D.C.: The National Academies Press, 2002).

[40] See: GAO, *Vehicle Fuel Economy: Reforming Fuel Economy Standards Could Help Reduce Oil Consumption by Cars and Light Trucks, and Other Options Could Complement These Standards*, GAO-07-921 (Washington, D.C.: Aug. 2, 2007). Other factors contributing to reduced transportation fuel consumption include higher fuel prices. See: National Academies of Sciences, 2002.

[41] See: EPA, *Final Rulemaking to Establish Light-Duty Vehicle Greenhouse Gas Emission Standards and Corporate Average Fuel Economy Standards Regulatory Impact Analysis*, EPA-420-R-10-009 (Washington, D.C.: April 2010).

[42] EPA uses renewable identification numbers to track compliance with the RFS. In this report, we refer to these as credits. Renewable fuel producers, refiners, and fuel blenders, among others, can acquire credits under the RFS, though only refiners and importers of petroleum-based transportation fuels are obligated under the RFS. If a company generates more credits than it needs for compliance in a given year, excess credits can be sold to others, or saved for the following compliance year, subject to certain constraints.

[43] See: GAO, *Biofuels: Challenges to the Transportation, Sale, and Use of Intermediate Ethanol Blends*, GAO-11-513 (Washington, D.C.: June 3, 2011), and *Biofuels: Potential Effects and Challenges of Required Increases in Production and Use*, GAO-09-446 (Washington, D.C.: Aug. 25, 2009).

[44] The RFS has also posed compliance challenges for refiners and other market participants because of concerns over fraud in the market for RFS compliance credits. Credits must be validly generated in order for renewable fuel producers to be able to transfer or use them for

compliance purposes. For example, EPA has issued several notices of violation alleging that five companies generated more than 170 million gallons of invalid credits without producing qualifying renewable fuels. These invalid credits had been used for compliance by many refining companies. EPA recently issued a Notice of Proposed Rulemaking, which describes a voluntary program known as the Quality Assurance Program to help assure that credits are validly generated. The notice proposes that voluntary third parties verify the validity of credits through steps such as verification of type of feedstocks, verification that volumes produced are consistent with amount of feedstocks processed, and verification that credits generated are appropriately categorized and match the volumes produced.

[45] The requirement for cellulosic biofuel began in calendar year 2010.

[46] For calendar years after 2012, EPA is also required to set biomass-based diesel volumes at least 14 months in advance. Though not shown in figure 9, EPA has also been late in issuing final standards for biomass-based diesel. EPA issued the final volume standards for 2013 in September 2012—about 10 months late. As of March 2014, EPA is 16 months late in finalizing standards for 2014 and 4 months late in finalizing standards for 2015. EPA proposed standards for 2014 and 2015 on Nov. 29, 2013, which have not been finalized.

[47] The RFS compliance deadline—the date by which refiners and other obligated parties must demonstrate compliance to EPA—is established by EPA through implementing regulations and has been adjusted by EPA. For example, EPA extended the compliance deadline for the 2013 compliance year from February 28, 2014 to June 30, 2014. The extension provided more time for refiners and other obligated parties to purchase and allocate credits.

[48] See GAO, *Federal Rulemaking: Improvements Needed to Monitoring and Evaluation of Rules Development as Well as to the Transparency of OMB Regulatory Reviews,* GAO-09-205 (Washington, D.C.: Apr. 20, 2009).

[49] These refineries are the Hyperion Refinery and Energy Center in South Dakota, the Sinclair Wyoming Refinery in Wyoming, and the Diamond Shamrock, Valero McKee Refinery in Texas.

[50] A major modification refers to a physical or operational change that would result in a significant net increase in emissions of a regulated pollutant. However, in our prior work, we noted that it can be difficult for unit owners and regulators to know whether an air pollution permit is needed, because permitting rules governing applicability are complex and because applicability is determined on a case-by-case basis. See: GAO, *Air Pollution: EPA Needs Better Information on New Source Review Permits,* GAO-12-590 (Washington, D.C.: June 22, 2012).

[51] State permitting agencies may also issue GHG permits to refineries subject to EPA guidance.

[52] In a consolidated lawsuit, a Midwest ethanol group and an association of petroleum refiners claimed that LCFS discriminates against petroleum and ethanol firms from outside the state, and that the regulation was preempted by the Clean Air Act. Following appeal of an adverse District Court decision, in September 2013, the Ninth Circuit Court of Appeals largely upheld the LCFS by remanding one issue back to the district court for trial. Groups representing the energy industry and farmers, among others, have petitioned for the case to be reheard before the full Ninth Circuit. LCFS was also challenged in state court on administrative procedural grounds, which we describe above. CARB officials told us that they were optimistic about the outcome of these legal challenges, and said that their expectation is that if the courts do find against CARB on any particular issues, they would be able to narrowly adjust the regulations in order to address the court's findings and did not anticipate needing to wholesale replace or rework the regulations.

[53] In December 2011, a federal judge granted a preliminary injunction against the implementation of the LCFS, among other rulings. CARB appealed the decision and the Ninth Circuit Court of Appeals stayed the injunction in April 2012 and later reversed the District Court on these rulings.

[54] The Boston Consulting Group, Understanding the impact of AB 32 (Boston, MA: June 2012).

[55] The IEA forecast that we reviewed did not include a projection of U.S. consumption. EIA data track "product supplied," which approximately represents consumption of petroleum products because it measures the disappearance of these products from sources such as refineries, bulk terminals, and pipelines. Because the future depends on a multitude of factors that can be difficult to predict, EIA's 2013 and initial 2014 forecast include 28 scenarios with different assumptions about future conditions, such as economic growth or policy changes. (See app. I for additional information on our methodology.)

[56] Forecasts with more recent data and revised assumptions can indicate somewhat different levels of consumption. For example, in its 2013 forecast, IHS projected higher domestic consumption in 2020 than its previous year's forecast. EIA recently released initial results of its 2014 forecast, in which EIA projected lower domestic consumption in the reference scenario compared with the reference scenario in the 2013 forecast also discussed in this report. EIA attributed the difference, in part, to reduced expectations for vehicles miles traveled.

[57] *The American Energy Initiative, Part 22: EPA Greenhouse Gas Regulations, Before the Subcomm. on Energy and Power of the H. Comm. on Energy and Commerce*, 112th Cong. (2012) (statement of EPA Assistant Administrator for Air and Radiation, Regina McCarthy). In addition, EPA specifically projected that the 2012-2016 CAFE and GHG vehicle emission standards would reduce domestic-refined gasoline consumption by about 7 billion gallons per year by 2020 and shrink domestic refining by about 3 percent by 2020. (See: EPA, *Final Rulemaking to Establish Light-Duty Vehicle Greenhouse Emission Standards and Corporate Average Fuel Economy Standards, Regulatory Impact Analysis* (Washington, D.C.: April 2010.)

[58] See EPA, *Renewable Fuel Standard Program (RFS2) Regulatory Impact Analysis*, EPA- 420-R-10-006 (Washington, D.C.: February 2010).

[59] Similarly, while EPA issued GHG vehicle emission standards through 2025, NHTSA is required to set CAFE standards only 5 years at a time. CAFE standards for model years 2022-2025 have been presented by NHTSA as "augural" standards in the interest of aiding manufacturers in future product planning, as well as harmonization. A midterm evaluation by NHTSA and EPA may result in modifications to both EPA standards and NHTSA's augural standards. In addition, as discussed below, future requirements under the LCFS—and their effect on refiners—are uncertain due to various factors.

[60] Hydrotreating removes sulfur by converting the sulfur into hydrogen sulfide gas and then to elemental sulfur.

[61] Baker & O'Brien Incorporated, *Addendum to Potential Supply and Cost Impacts of Lower Sulfur, Lower RVP Gasoline* (March 2012).

[62] Turner, Mason & Company, *Economic and Supply Impacts of a Reduced Cap on Gasoline Sulfur Content* (Dallas, Texas. February 2013).

[63] See: Settlement Agreement, *State of New York, et al, v. Environmental Protection Agency*, No. 08-1279 (D.C. Cir. Dec. 29, 2010).

[64] California Air Resources Board, *Low Carbon Fuel Standard 2011 Program Review Report* (Sacramento, CA: December 2011).

[65] Boston Consulting Group, *Understanding the Impact of AB32* (Boston, MA: June, 2012).

[66] For example, an electric car could receive three additional credits, thus reducing the need for low carbon fuels with limited availability.

[67] Unless noted otherwise, exports of petroleum products in this section refer specifically to finished petroleum products, natural gas liquids and liquid refinery gases, oxygenates (excluding fuel ethanol), and motor gasoline blending components. Finished products include finished motor gasoline, finished aviation gasoline, kerosene-type jet fuel, kerosene, distillate fuel oil, residual fuel oil, and other products.

[68] According to EIA, U.S. exports of diesel reached a record monthly high in July 2013, at 1.4 million barrels per day, and remained strong in August and September 2013. Continued growth in global diesel consumption has supported U.S. exports.

[69] We did not include IHS data in this figure because they are not public.

[70] Excess capacity may result in some refinery closures. IEA's New Policies Scenario projects 6.3 million barrels per day of global refining capacity may be at risk of closure in 2020, but the projected capacity at risk is outside of North America.

[71] See International Energy Agency, *World Energy Outlook 2013* (Paris, France: November 2013).

[72] According to EPA, Japan, the 30 nations that comprise the European Union, Albania, and Bosnia-Herzegovina have 10 ppm gasoline sulfur caps in place.

[73] Closely associated with this trend has been the increasing availability and lower costs of natural gas, which has also provided a benefit to U.S. refiners.

[74] EIA expects most of the increase to come from onshore tight oil formations in the Bakken, Eagle Ford, and Permian Basin regions, though increases are likely in domestic offshore production as well.

[75] EIA's most recent forecast, the Annual Energy Outlook 2014 Early Release, contains reference scenario projections only. Projections for other scenarios were not available at the time of this report's publication.

[76] According to EIA, petroleum liquids include crude oil and lease condensate as well as natural gas plant liquids, bitumen, extra-heavy oil, and refinery gains.

[77] For example, in July 2013, a deadly explosion occurred after a railcar carrying crude oil derailed in Quebec. In response to petitions it received following the accident, DOT's Pipeline and Hazardous Materials Safety Administration is considering whether to revise its regulations to improve railcar safety standards. See 78 Fed. Reg. 54849 (Sept. 6, 2013).

[78] According to EIA data, total crude oil exports totaled 67,000 barrels per day in 2012—1 percent of total 2012 production. U.S. exports of crude oil must be licensed by the Department of Commerce, Bureau of Industry and Security. Current regulations specify limited conditions in which license applications will be approved, for example, for exports to Canada for use or consumption in Canada and that are transported other than through the Trans-Alaska Pipeline, for which some restrictions apply. Other applications for export licenses are to be reviewed on a case-by-case basis and generally will be approved if determined to be consistent with the national interest and the purposes of the Energy Policy and Conservation Act, and any additional statutory controls are satisfied. See generally 15 C.F.R. § 754.2 (2013).

[79] Department of Energy, Office of Policy and International Affairs, *Small Refinery Exemption Study: An Investigation into Disproportionate Economic Hardship*, March, 2011.

[80] Department of Energy, Energy Information Administration, *Annual Energy Outlook 2013 with Projections to 2040*, DOE/EIA-0383(2013) (Washington, D.C.: April 2013); Department of Energy, Energy Information Administration, *Annual Energy Outlook 2014 with Projections to 2040, Early Release*, DOE/EIA-0383ER(2014) (Washington, D.C.: December 2013); International Energy Agency, *World Energy Outlook 2013* (Paris, France: November 2013); IHS, *US Energy Outlook 2013* (Englewood, CO: November 2013).

[81] EIA's 2014 Annual Energy Outlook Early Release contains reference scenario projections only. Other scenarios were not available at the time of this report's publication.

[82] We highlight the following scenarios in the report: reference (2014), low or no net imports, high net imports, high economic growth, low oil and gas resource, greenhouse gas $25 and low gas prices, and greenhouse gas $10 and low gas prices.

[83] We also obtained views from officials at the Energy Information Administration within the Department of Energy.

[84] We have previously reported on various aspects of the blend wall and other challenges to the increased use of renewable fuels. See GAO-11-513 and GAO-09-446.

[85] The RFS requirements are "nested" in that renewable fuels other than ethanol could be used to meet the total renewable fuels requirement.

In: U.S. Petroleum Refining Industry … ISBN: 978-1-63321-472-9
Editor: Collin T. Roberts

Chapter 2

ENERGY MARKETS: INCREASING GLOBALIZATION OF PETROLEUM PRODUCTS MARKETS, TIGHTENING REFINING DEMAND AND SUPPLY BALANCE, AND OTHER TRENDS HAVE IMPLICATIONS FOR U.S. ENERGY SUPPLY, PRICES, AND PRICE VOLATILITY*

United States Government Accountability Office

WHY GAO DID THIS STUDY

To better understand how changes in domestic and international petroleum products markets have affected prices, GAO was asked to evaluate trends in (1) the international trade of petroleum products, (2) refining capacity and intensity of refining capacity use internationally and in the United States, (3) international and domestic crude oil and petroleum product inventories, and (4) domestic petroleum supply infrastructure.

To address these objectives, we reviewed numerous studies, evaluated data, and spoke to many industry officials and experts and agency officials.

* This is an edited, reformatted and augmented version of the United States Government Accountability Office publication, GAO-08-14, dated December 2007.

What GAO Recommends

GAO is making recommendations aimed at improving the functioning of petroleum product markets, including that the Secretaries of Transportation and Energy coordinate with other agencies to (1) encourage more uniform biofuel and petroleum product blending practices, (2) conduct a study of infrastructure system adequacy, and (3) evaluate the assignment of a lead agency to coordinate permitting of infrastructure construction.

In commenting on the report, the Federal Energy Regulatory Commission generally agreed with the report's findings and recommendations, while the Departments of Energy and Transportation neither fully agreed nor disagreed.

What GAO Found

International trade in petroleum products has expanded over the past two decades, making markets for gasoline and other petroleum products increasingly global in nature. Recent plans and mandates in the United States and other countries to greatly expand the use of biofuels blended with petroleum products—for example, ethanol blended with gasoline and biodiesel blended with petroleum diesel—may have the unintended effect of reducing opportunities for trade because blending different levels of biofuels with petroleum blending stocks will require changes to these blending stocks and thereby reduce their fungibility.

For most of the past 25 years, there has been excess refining capacity globally, but this excess has shrunk considerably in recent years as demand has increased faster than capacity growth, causing refineries to run closer to their production capacity, and contributing to recent increases in petroleum product prices, price volatility, and refining profits. However, experts say it is unclear whether or for how long the current market tightness will continue, in part because of uncertainties about how much additional refining capacity will actually be built in the face of rising construction costs and initiatives that may reduce future demand for petroleum products such as through the blending of large volumes of biofuels into the transportation fuels markets.

When measured as average days of consumption, inventories of petroleum products and crude oil in the United States indicate a general decline over the past 20 years. A number of factors have contributed to this decrease in the United States, including reductions in crude oil production and the number of refineries as well as efforts to reduce inventory holding costs by applying advances in technology. Lower operating costs associated with lower inventories may have translated into lower consumer prices during normal periods. However, lower than normal inventories can lead to higher or more volatile prices in the event of supply disruptions or surges in demand.

The nation's petroleum product supply infrastructure is constrained in key areas and is likely to become increasingly constrained, unless timely investments are made. A constrained supply infrastructure can exacerbate price effects and price volatility due to a supply disruption. However, no central source of data tracks system bottlenecks. While there is widespread recognition that a study is needed to fully identify the extent of infrastructure inadequacy and the impact on prices, to date, no such analysis has been undertaken, though such a study was mandated by Congress in 2006 with a June 2008 deadline. Significant infrastructure expansion plans in the private sector could alleviate the stresses. However, a complex permitting and siting process involving as many as 11 federal agencies and numerous state and local stakeholders has slowed or impeded the expansion and construction of new pipelines. Unlike in the case of natural gas pipelines, no central federal agency acts to coordinate this permitting process.

ABBREVIATIONS

BP	British Petroleum
DOE	Department of Energy
DOT	Department of Transportation
EIA	Energy Information Administration
EPA	Environmental Protection Agency
FERC	Federal Energy Regulatory Commission
FTC	Federal Trade Commission
IEA	International Energy Agency
MARAD	U.S. Maritime Administration
NYMEX	New York Mercantile Exchange
OECD	Organisation for Economic Co-operation and Development
SPR	U.S. Strategic Petroleum Reserve

December 20, 2007

The Honorable Daniel Inouye
Chairman, Committee on
Commerce, Science, and Transportation
United States Senate

The Honorable Maria Cantwell
United States Senate

In 2003, the price of West Texas Intermediate crude oil, a widely watched benchmark crude oil price, averaged about $31 per barrel. By 2006, the average was about $66 per barrel, and in mid-November, 2007 the price rose to over $90 per barrel. Wholesale and retail prices of petroleum products refined from crude oil, including gasoline, diesel, and jet fuel, which normally rise and fall with crude oil prices, also generally rose over the period. For example, U.S. retail regular gasoline prices—equivalent to wholesale prices plus taxes, marketing costs, and retail profit margins— averaged $1.52 per gallon in 2003, but by August 2006, they had almost doubled to $3.00 per gallon, and as of July 2007, remained relatively high at $2.85 per gallon. Such large and sustained increases in gasoline prices have not been seen in the United States since the late 1970s and early 1980s, when the start of the Iran-Iraq war pushed prices up—even higher than today's prices when adjusted for inflation—causing severe economic hardship for many Americans and contributing to a global economic recession. While this more recent increase in petroleum product prices does not appear to have had such far-reaching economic effects, consumers want to know the reasons for the large and relatively sudden price increases. Figure 1 shows retail regular gasoline prices in the United States, in both nominal and inflation-adjusted terms during the past 30 years.

In addition to crude oil prices, a number of factors affect the price of petroleum products. As we recently testified before Congress, these factors include domestic capacity to refine crude oil into petroleum products; inventories of these products; the proliferation of special blends of gasoline; the capacity and functioning of the crude oil and petroleum product supply infrastructure, which is composed of pipelines, barges, tanker vessels, marine terminals, rail, trucking and storage tanks; and mergers in the oil industry.[1] In addition, because the United States imports and exports petroleum products, events outside the United States can affect domestic petroleum product prices.

Imports to or exports from the United States typically enter or leave through port facilities on tankers or across national borders via pipeline. Our imports of petroleum products come from all over the world into ports in the Gulf of Mexico and the east and west coasts, and by pipeline from Canada.

Refineries process crude oil into petroleum products through a variety of complicated processes, and a single barrel of crude oil produces a varying amount of gasoline, diesel, jet fuel, and other products depending on the configuration—or complexity—of the refinery as well as the type of crude oil being refined. Refineries can be optimized—or "upgraded"—to process different grades of crude oil through the addition of specialized refining equipment. U.S. refineries are generally optimized to produce large proportions of gasoline to meet domestic transportation demand. Cleaner-burning fuels have proliferated in response to legislation including the Clean Air Act Amendments of 1990, leading to additional investments in the refining equipment needed to produce the new fuels.

More recently, a number of European countries, the U.S. federal government, and a number of individual states and localities have proposed or mandated the use of biofuels—such as ethanol made from corn or biodiesel made from soybeans or other crops—partly in an effort to reduce greenhouse gas emissions and reduce consumption of petroleum products. These mandates call for biofuels to be blended in varying proportions with traditional gasoline or diesel. For example, U.S. federal biofuel standards call for a minimum proportion and volume of biofuels to be sold each year but do not specify how that proportion is met. In addition, a number of states and at least one city have requirements or plans to require use of biofuels in varying proportions, blended with gasoline and diesel. For example:

- Hawaii, Minnesota, and the city of Portland, Oregon, all currently require ethanol to be blended at a 10 percent by volume rate with gasoline, although Hawaii only requires this for 85 percent of the gasoline sold in the state.
- Minnesota and Portland, Oregon require 2 and 5 percent biodiesel, respectively, to be blended with diesel fuel. Minnesota also requires the expansion of ethanol blending to 20 percent by volume by 2013.
- Four other states—Missouri, Montana, New Mexico, and Oregon—have biofuel mandates that will require 10 percent ethanol blended into gasoline and/or varying blends of biodiesel: Missouri and Montana have no mandated plans for biodiesel; New Mexico calls for 5 percent biodiesel blending and Oregon for 2 percent.

- Other states have "flexible standards." For example, Iowa provides tax credits if at least 10 percent of the fuels used by 2009 are renewable, with the threshold rising to 23 percent in 2018. Yet this can be achieved in a flexible way, using a blend consisting of 85 percent ethanol and 15 percent gasoline, while other gasoline would be blended with less or no ethanol at all. Louisiana will require both ethanol and biodiesel to be blended at 2 percent, but only when state production reaches certain levels and prices of ethanol and biodiesel are sufficiently low. Finally, Washington will require that at least 2 percent of diesel sold be biodiesel by November 30, 2008, or when a determination is made that state biodiesel production can meet the 2 percent requirement.

Automakers and refiners told us that these varying biofuel blends will require changes to the gasoline and diesel blendstocks—the fuels that will be mixed with the ethanol or biodiesel—to maintain engine performance and emissions requirements. The production of these new blends may also require further refinery changes as well as changes to automobile engines. Automakers also told us that in addition to increasing the costs of production, changing engines to be able to meet performance and emissions standards using a wide mix of biofuel blends would also entail potential losses in fuel efficiency.

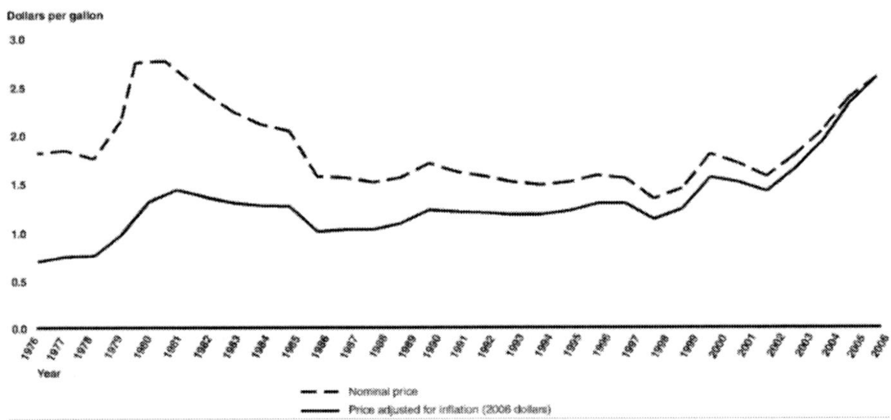

Source: GAO analysis of EIA data.

Figure 1. U.S. Retail Regular Unleaded Gasoline Prices, Annual Average, 1976 - 2006.

From refineries, petroleum products are distributed through an extensive supply infrastructure composed of pipelines, barges, tanker vessels, marine terminals, rail roads, trucks, and storage tanks. Pipelines are generally the cheapest domestic mode for transporting crude oil and petroleum products. Crude oil and petroleum products are transported in separate pipelines, and while different types and specifications of petroleum products are shipped in the same pipelines, they must be kept separate during transport and storage in order to maintain the specific desirable performance and emissions characteristics of these different fuels. Crude oil pipelines connect several large refining centers to crude oil sources, and petroleum product pipelines connect these refineries to population centers all over the country. Trucks and rail have generally distributed only a small fraction of petroleum products to wholesale terminals. However, they are being increasingly utilized to move ethanol to locations near final demand centers where the ethanol is blended with gasoline. This is because existing pipelines cannot currently accommodate ethanol due to an insufficient collector pipeline network linking ethanol refineries with major pipelines, and because ethanol has corrosive and other properties that complicate its transport in pipelines that also carry petroleum products.

Refiners, distributors, and marketers of petroleum products maintain inventories of crude oil and petroleum products to facilitate smooth supply operations and mitigate the effects of supply disruptions. Crude oil and petroleum product inventories consist of three levels. Primary inventories comprise the crude oil or petroleum products held at production sites, refineries, and storage terminals, and in pipelines, tankers, barges, and other transportation centers. Secondary inventories consist of retail outlets and small storage facilities—those with less than 50,000 barrels of total capacity—that exist between the primary distribution system and the end user. Tertiary inventories are the petroleum products in the hands of end users, for example, in drivers' gasoline tanks. The federal government also maintains strategic stocks of crude oil and, in the Northeast, heating oil to be released in the event of a major supply shortage. The Energy Information Administration (EIA) collects inventory data for the primary system. Information about changes in inventory levels can inform market participants about underlying demand or supply conditions that will influence prices.

A number of federal agencies have programs and activities related to the oversight or monitoring of the refining, distribution, or importing of petroleum and petroleum products. For example, the Department of Transportation (DOT) oversees crude oil and petroleum product pipelines and monitors their operations to ensure public safety. The Federal Energy Regulatory Commission (FERC) regulates the "tariffs"—or rates and conditions—under which interstate crude oil and petroleum product pipelines operate, while individual states have oversight over intrastate pipelines within their borders. EIA collects data from refiners and others about shipments of crude oil and petroleum products by pipeline and barge between regions of the United States. In addition, a number of federal and state agencies and other local and private entities become involved in approving new supply infrastructure projects. For example, the approval to build or repair a pipeline could involve DOT's Pipeline and Hazardous Materials Safety Administration, the Environmental Protection Agency, Bureau of Land Management (if pipelines cross federal lands), Army Corps of Engineers, U.S. Fish and Wildlife Service, as well as other federal agencies, and state and local stakeholders. Legislation in 2002 mandated the formation of an interagency committee to help expedite pipeline review and permitting processes for pipeline repairs. That committee is composed of 11 federal entities. For construction of interstate natural gas pipelines, the Federal Energy Regulatory Commission takes a lead role in coordinating the permitting process across the relevant federal agencies and can convey the right of eminent domain to builders of natural gas pipelines to resolve disputes with owners of land needed to build a pipeline.[2] However, no such federal coordinating authority or power of eminent domain exists for construction and expansion of new interstate petroleum product or crude oil pipelines.

To better understand changes in domestic and international markets for petroleum products and the implications of these changes for recent price increases, you asked us to evaluate trends and effects on petroleum product prices in (1) international trade of petroleum products; (2) refining capacity and intensity of refining capacity use internationally and in the United States; (3) international and domestic crude oil and petroleum product inventories; and (4) domestic crude oil and petroleum product supply infrastructure, particularly pipelines and marine transportation.

To evaluate trends in the international trade of crude oil and petroleum products, we analyzed data from EIA and the International Energy Association (IEA) and spoke with numerous government agency and oil company officials and industry experts. To assess trends in refining capacity, we evaluated IEA, EIA, and *Oil and Gas Journal* data, and spoke with numerous industry experts. To evaluate trends in inventories of crude oil and petroleum products, we reviewed data from EIA and IEA on inventories and demand to analyze international, U.S. national, and regional inventories. We analyzed New York

Mercantile Exchange (NYMEX) and other futures market data, as well as EIA data, and asked experts about the effects of these futures prices for crude oil on inventory holding decisions. To evaluate trends in supply infrastructure for crude oil and petroleum products, we collected and analyzed available data on the pipeline and marine infrastructure system, capacity, throughputs, and constraints. We examined reports and data from supply disruption case studies to examine those cases' impact on prices and price volatility. We spoke with numerous government agency and pipeline company officials and industry experts. This report focuses on long-term trends in the industry, rather than recent events that have influenced prices of gasoline and other petroleum products. GAO currently has ongoing work looking at such recent trends as refinery outages through the spring of 2007 and mergers in the industry since 2000.

This report uses data from domestic and international wholesale petroleum products and crude oil markets and domestic retail markets. In contrast to retail prices, wholesale prices do not include taxes, distribution and marketing expenses, and profits. In every case for the data used in this report, we assessed and determined that the data were sufficiently reliable for our purposes. A more detailed description of the scope and methodology of our review is presented in appendix I. We performed our work from August 2006 through September 2007 in accordance with generally accepted government auditing standards.

RESULTS IN BRIEF

International trade in petroleum products has expanded significantly over the past two decades, making markets for gasoline and other petroleum products increasingly global in nature. This trend has been particularly important for the United States; while in 1970 the United States was largely self-sufficient in gasoline, we now import over 10 percent of our annual gasoline consumption. Having access to more sources of supply can benefit the United States in the event of domestic supply disruptions. For example, the benefit of such flexibility in sources of supply helped U.S. marketers and retail sellers obtain gasoline and other petroleum products in the aftermath of Hurricanes Katrina and Rita, when imports of gasoline to the United States increased to fill the void left by damaged or shut-down domestic refineries and pipelines. However, the fact that petroleum product markets are international means that supply disruptions or unexpected increases in demand anywhere in the world can influence U.S. prices. Our analysis of wholesale prices in the United States, Europe, and Asia shows that prices in geographically dispersed markets rose significantly following Hurricanes Katrina and Rita, indicating that prices in these markets are linked to some extent. We also evaluated petroleum product import data and found that products came from a wider range of countries during this period, again indicating that products move in response to price signals globally. Recent plans and mandates in the United States and other countries to greatly expand the use of biofuels blended with petroleum products—for example, ethanol blended with gasoline and biodiesel blended with petroleum diesel—may have the unintended effect of reducing opportunities for trade because blending different levels of biofuels with petroleum blending stocks will require changes to these blending stocks and thereby reduce their fungibility. For example, if European countries adopt widely different blending levels biofuels in gasoline and diesel products as current plans call for, the refineries serving these countries will have to alter petroleum blending stocks for those blending levels, and this could make the blending stocks themselves less tradable across countries.

For most of the past 25 years, there has been excess refining capacity globally, but this excess capacity has shrunk considerably in recent years as demand has increased faster than capacity growth, causing refineries to run closer to their production capacity, and, along with rising crude oil prices, contributing to recent increases in petroleum product prices and price volatility. Demand for petroleum products has grown more quickly than has refinery capacity for much of the past 25 years, in large part because excess refining capacity historically caused profitability of the refining sector to be low compared to that of many other industries. More recently, this tightening of the balance between supply and demand for petroleum products has, along with higher crude oil prices and other factors, contributed to increased petroleum product prices and higher oil industry profits, and could contribute to greater price volatility. Recently high petroleum product prices and increased profits over those seen during the 1990s in the refining industry have spurred new refinery capacity investments in the United States and internationally. However, experts say it is unclear whether or for how long the current market tightness will continue, in part because of uncertainties about how much additional refining capacity will actually be built in the face of rising construction costs, and initiatives that may reduce future demand for petroleum products such as through the blending of large volumes of biofuels into the transportation fuels markets in many countries. The absence of national standards for blending biofuels with gasoline and diesel could also increase the number of gasoline and diesel blending stocks refiners have to make, which could require additional refining investment to make those blends that could crowd out investment in refining capacity expansions.

When measured as average days of consumption, long-term trends in inventories of petroleum products and crude oil in the United States indicate a general decline over the past 20 years. Similarly, gasoline and crude oil inventories in the Organisation for Economic Co-operation and Development (OECD) countries, excluding the United States, have also generally fallen over the same period.[3] Inventories, as measured by EIA IEA, and others, have some limitations as a measure of what is available to meet demand in the event of a supply shortfall, in part because the United States has imported an increasing share of its gasoline over the period during which inventories have fallen, and as such, the domestic inventory data do not account for large volumes of these products on the water or in tankers from foreign sources that are destined for the U.S. market or in storage terminals at foreign ports serving this trade in gasoline. A number of factors have contributed to the long-term decrease in inventory holdings in the United States, including reductions in both domestic crude oil production and the number of refineries. Advances in technology and changes in management processes also may have contributed to reduced inventories by enabling refiners to more closely time the production of supplies to meet expected demand. Lower operating costs associated with lower inventories may have translated into lower consumer prices during normal periods. However, in the short term, because inventories provide a smoothing effect against temporary demand and supply fluctuations, lower than normal inventories can lead to higher or more volatile prices in the event of supply disruptions or surges in demand.

The nation's petroleum product supply infrastructure is constrained in key areas and is likely to become increasingly constrained, unless timely investments are made. Industry and federal agency officials report a systemic lack of pipeline capacity in the supply infrastructure system in key states including Arizona, California, Colorado, and Nevada, and note the existing supply infrastructure is insufficient to carry the commensurate volumes of petroleum products and crude oil needed to meet growing demand there. A constrained supply infrastructure can exacerbate price effects and price volatility due to a supply disruption. For example, during a pipeline outage in 2003 that affected pipeline supplies to Arizona, retail prices of gasoline rose by about 45 cents per gallon. However, we were unable to assess the full extent of supply infrastructure constraints or the impacts of these constraints on prices and price volatility, in large part because there is no central source of data that tracks system bottlenecks. In 2006, DOT put forth a legislative proposal and Congress passed legislation that mandated the Secretaries of Energy and Transportation to conduct periodic analyses of (1) where unplanned petroleum product pipeline outages or insufficient pipeline capacity increase prices and (2) whether or not regulation is adequate to minimize the potential for unplanned losses of pipeline capacity. While there is widespread recognition that such a study is needed to fully identify the extent of infrastructure inadequacy and the impact on prices, to date, no such analysis has been undertaken. DOT and Department of Energy (DOE) officials told us that they were not allocated funds specifically to do the mandated analyses and that the agencies have not re-allocated other funds for this, although DOE told us it has met with DOT to discuss how this work could be approached. However, given that the study has not begun, it seems highly unlikely that agencies will be able to meet their June 2008 deadline for reporting to Congress. There are many private sector plans to expand the supply infrastructure, and if implemented in a timely fashion, these plans could significantly alleviate the stresses on the system. However, a complex permitting and siting process involving as many as 11 federal agencies and numerous state and local stakeholders has slowed or impeded the expansion and construction of new pipelines. The permitting process for building natural gas pipeline infrastructure has been made easier by the designation of FERC as a lead federal agency to streamline permitting for interstate natural gas pipeline expansion, but no such lead federal agency exists to facilitate permitting of crude oil or petroleum product pipeline construction or upgrading.

GAO is making recommendations aimed at improving the functioning of petroleum product markets, including that the Secretaries of Transportation and Energy coordinate with other relevant agencies to (1) encourage uniform biofuel and petroleum product blending practices, (2) conduct a study of infrastructure system adequacy, and (3) evaluate the feasibility of assigning a lead federal agency to coordinate the permitting of infrastructure construction. In commenting on the report, the Federal Energy Regulatory Commission generally agreed with the report's findings and recommendations, while the Departments of Energy and Transportation neither fully agreed nor disagreed.

BACKGROUND

The United States is the largest consumer of crude oil and petroleum products of all nations, though demand for crude oil is growing faster globally, led by growth in developing countries such as China and India. When processed, crude oil is refined to produce petroleum products such as gasoline, diesel, and jet fuel, which have been instrumental in providing the nation with affordable fuel for automobiles, trucks, airplanes, and other forms of transportation and heating in some parts of the country. The petroleum industry consists of three main segments: the exploration and production segment (upstream); the refining and marketing segment (downstream); and a third segment typically referred to as the midstream, which consists of the infrastructure used to transport crude oil and petroleum products. Several U.S. agencies regulate and monitor the downstream and midstream oil industry and petroleum product markets.

The United States Is the Largest Consumer of Crude Oil and Petroleum Products, but Global Demand Has Grown Significantly in Recent Years

While the United States is the largest consumer of crude oil and petroleum products, global demand for crude oil and petroleum products is growing at a faster pace than U.S. demand, driven by growing consumption of crude oil and certain petroleum products in developing countries such as China and India. In 2006, the United States' share of world oil consumption was approximately 25 percent. The EIA projects in its reference, or "baseline," scenario that world oil consumption will continue to grow and will reach 118 million barrels per day in 2030. About 43 percent of this growth will come from non-OECD countries, particularly China and India, but the United States will remain the world's largest consumer. Under the assumptions of EIA's reference case scenario, U.S. demand for oil is projected to increase by 30 percent between 2005 and 2030—from about 21 million barrels per day in 2005 to about 27 million barrels per day in 2030—compared to 39 percent for the entire world. Meanwhile, domestic production of oil has generally been in decline for decades, leading to greater reliance on imported oil. In 2006, the United States imported about 66 percent of its crude oil.

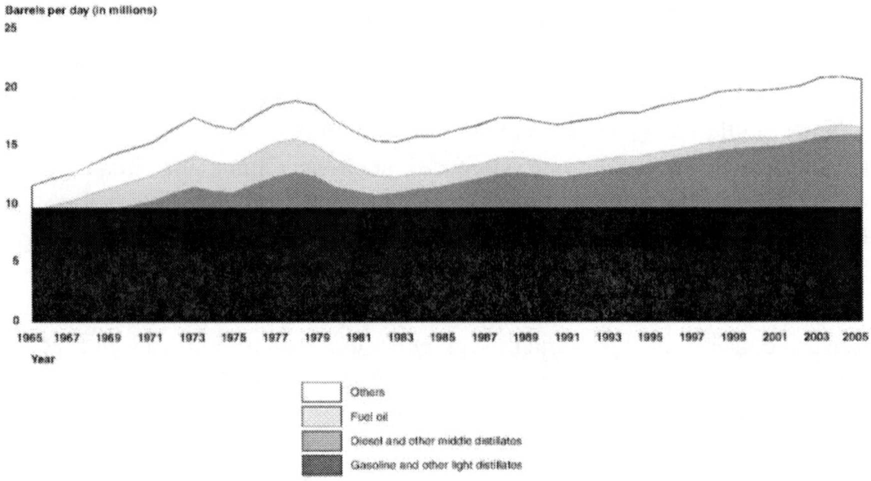

Source: GAO analysis of BP Statistical Review of World Energy June 2007.
Notes:
Other light distillates consists of aviation gasoline and light distillate feedstock (LDF)
Other middle distillates consists of jet and heating kerosene, and gas and diesel oils (including marine bunkers)
Fuel oil includes marine bunkers and crude oil used directly as fuel
Others consists of refinery gas, liquid petroleum gas, solvents, petroleum coke, lubricants, bitumen, wax, and other petroleum products and refinery fuel and loss

Figure 2. U.S. Consumption of Petroleum Products, 1965-2006.

When processed, crude oil produces petroleum products such as gasoline, diesel, and jet fuel, which have been instrumental in providing the nation with affordable fuel for automobiles, trucks, airplanes, and other forms of transportation, and—in some parts of the country—heating. Overall, demand for petroleum products in the United States has generally increased over the last 25 years, as shown in figure 2.

Key Aspects of the Petroleum Product Markets: Refining, Inventories, and Infrastructure

The petroleum industry consists of three main segments: the exploration and production segment (upstream); the refining and marketing segment (downstream); and a third segment typically referred to as the midstream, which consists of the infrastructure used to transport crude oil and petroleum products. This report is mainly concerned with certain aspects of the downstream and midstream segments, namely refining, inventories, and the pipeline and marine supply infrastructure.

Refining

Refineries change crude oil into petroleum products primarily through a distillation process that separates the crude oil into different fractions based on boiling point ranges. One barrel of crude oil produces a varying amount of gasoline, diesel, jet fuel, and other petroleum products depending on the configuration–or complexity–of the refinery and the type of crude oil that is being refined. Through the addition of specialized equipment, refineries can be optimized—or "upgraded"—to produce greater proportions of specific types of products or to use different grades of crude oil. For example, hydrocracking units enable refiners to increase the production of lighter fuels including gasoline, diesel fuel, and jet fuel; catalytic cracking units increase the production of gasoline; and hydrotreating units enable refiners to produce lower-sulfur fuels required by the European Union, United States, and many other countries. Changes in product specifications, shifts in demand, and environmental regulations all have important implications for refineries. For example, the regulated shift to unleaded gasoline that began in 1974 caused refineries to install equipment to produce high-octane components to replace the lost lead. Similarly, in response to environmental regulations such as limits on the emissions of certain air pollutants refineries have invested in equipment and processes to control such emissions. The proliferation of some special gasoline blends, or "boutique fuels," has made it more complicated to supply gasoline and raised costs, significantly affecting operations at refineries.[4] Last, to the extent that varying amounts of biofuels blended with gasoline and diesel require changes to the gasoline and fuel blendstocks, further refinery changes may be required to accommodate these blends. Shifting demand for petroleum products, such as Europe's declining demand for gasoline and growing consumption of diesel, can also cause refiners to invest in different processes to produce the mix of products desired by the market.

In general, the United States' refineries are among the most sophisticated in the world, and domestic refineries have generally been optimized to produce large proportions of cleaner-burning gasoline to meet the huge transportation demand subject to various environmental constraints. Historically, U.S. and international refining capacity has broadly grown and fallen in response to shifts in demand for petroleum products. For example, U.S. and international refining capacity fell sharply during the early 1980s in response to falling demand for petroleum products, caused in part by high prices of these products and worldwide recession. By 1983, demand had fallen so much that almost 30 percent of U.S. refinery capacity was not being used. Many refineries were shut down or idled and refining capacity thus fell. Demand began growing again in the United States and internationally around 1982.

Inventories

Inventories of petroleum products and crude oil are maintained by refiners, distributors, marketers, and others to mitigate the effects of disruptions, and to ensure a continuity of supply to their customers. Companies build inventories in preparation for planned maintenance and production, refining, and logistical systems. The primary inventory system comprises the crude oil or petroleum products held at production sites, refineries, and storage sites, and in pipelines, tankers, barges, and other transportation centers. Secondary inventories exist between the primary distribution system and the end user, and consist of retail outlets and small tank farms, which have less than 50,000 barrels of total capacity. Tertiary inventories are inventories held by consumers, for example, in automobile tanks. EIA collects inventory data for the primary system. EIA collects inventory data for crude oil and petroleum products held in storage at refineries, pipelines, and tank farms, and bulk terminals that can store at least 50,000 barrels of petroleum products. EIA also collects inventory data for Alaskan crude oil in transit by tanker from the terminus of the Alaskan pipeline in Valdez, Alaska, to other U.S. ports, as well as oil in the Strategic Petroleum Reserve.[5]

Pipeline and Marine Supply Infrastructure System

The supply infrastructure is composed of petroleum product and crude oil pipelines, barges, vessels, marine terminals, and storage tanks. Trucks and rail also distribute a small fraction of the products—about 6 and 4 percent respectively—but are being increasingly utilized with the rise of biofuels such as ethanol, which existing pipelines cannot currently accommodate.[6]

As displayed in figure 3, about 90 percent of all petroleum products currently travel by either pipeline or marine transport. Because of these two modes dominance, our analysis of the nation's supply infrastructure system will be limited to the pipeline and the marine transport system.

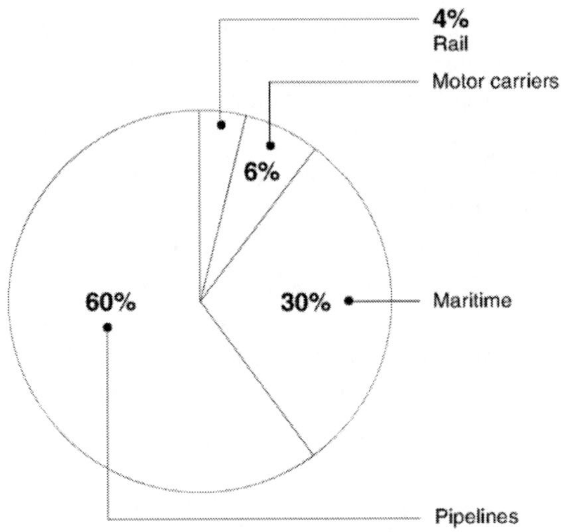

Source: GAO analysis of Bureau of Transportation Statistics data.

Figure 3. Transport Mode of Petroleum Products in the United States, 2004.

Pipelines are generally the least expensive mode for transporting oil and most petroleum products.[7] Most of the United States pipeline infrastructure—approximately 166,000 miles of crude oil and petroleum product pipeline—was constructed in the 1950s, 1960s and 1970s to accommodate the needs of the refining sector and demand centers at that time. These main pipelines were built to transport petroleum products from the Gulf Coast and Midwestern oil fields—where many of the nation's refineries were—to the East Coast, the United States' largest consuming region. The first large transmission pipelines for petroleum products were constructed during World War II, and ran primarily from the Gulf Coast to the Mid-Atlantic states; the network expanded significantly until the 1970s. Pipelines feed refinery centers and market hubs because the regions with the most supply are not the regions with the most demand. Most pipelines are common carriers, offering transportation services to anyone who wants them, but subject to some regulations. While crude oil and petroleum products generally do not travel on the same pipelines, numerous different petroleum products are shipped back to back in batches through the same pipelines. During this process, some blending of any two adjacent batches of petroleum products occurs where the two batches interface. This blended material may be simply mixed with the lower-valued product—for example, the mix of high- and low- octane gasoline at the interface between batches of these commodities would be downgraded, or mixed with the low-octane fuel—or, if the blended material is incompatible with either of the two petroleum products that interfaced, it must be removed and reprocessed into something that can be used. To access space on a pipeline, a shipper must ask for the right to use capacity by nominating amounts of liquid for service to be received, delivered or stored by the pipeline company. Different shippers' nominations of common products are often combined by the pipeline in order to reduce the number of batches and therefore the amount of downgrading or reprocessing of blended products.

Marine transportation of crude oil and petroleum products accounts for nearly one-third of domestic shipments. The marine transport system consists primarily of waterways; ports and vessels, including crude oil tankers; and product tankers and tank barges. Built to accommodate smaller vessels, many of the major ports have had to expand in response to increasing marine transport and trade and to accommodate larger tanker vessels.

Several U.S. Agencies Regulate and Monitor the Downstream and Midstream Oil Industry and Petroleum Product Markets

Several U.S. agencies have jurisdiction over or monitor the U.S. downstream oil industry and petroleum product markets:

- Within DOE, EIA collects and analyzes data on the supply, consumption, and prices of crude oil and petroleum products, including inventory levels, refining capacity and utilization rates, and product movements into and within the United States. DOE's Office of Fossil Energy manages the U.S. Strategic Petroleum Reserve (SPR), which is a federally maintained stockpile of about 700 million barrels of crude for use in the case of a major disruption of oil supplies, as well as the Northeast Home Heating Oil Reserve, a component of the SPR that has 2 million barrels of emergency fuel oil for homes and businesses in the Northeast that could be released during heating oil supply interruptions or high periods of demand caused by severe winter weather.
- FERC is an independent agency that regulates the transmission of oil through interstate pipelines by setting and enforcing pipeline "tariffs"—the prices and terms under which shippers send their

products through the pipelines and the rules governing access to these pipelines.[8]

- The Federal Trade Commission (FTC) enforces antitrust and consumer protection statutes. For example, in the petroleum industry, the FTC generally reviews proposed mergers and approves such mergers only if they are deemed not to have anticompetitive effects.
- DOT's Pipeline and Hazardous Materials Safety Administration regulates safety for oil pipelines that transport oil and petroleum products. Among other things, it oversees oil pipelines' design, maintenance, and operating procedures. DOT's Maritime Administration (MARAD) reports to Congress on the status of public ports' supply infrastructure needs.
- The Environmental Protection Agency (EPA) develops and enforces regulations that implement environmental laws including the Clean Air Act, the Clean Water Act, and the Oil Pollution Act, which aim to control the discharge of pollutants into the environment by refiners and other industries. EPA also administers the National Environmental Policy Act, which requires federal agencies to consider environmental impacts of proposed actions.

In addition, individual foreign countries play regulatory roles and can affect trade conditions for products through their individual or collective actions. The IEA is an organization established by treaty of 26 mainly net oil-importing OECD countries to cope with oil supply disruptions and coordinate an international response in the case of a disruption to the global oil supply. Member countries agree to keep significant strategic stocks of crude oil and/or petroleum products to be available in the event of a severe supply disruption. IEA also maintains a database that provides information on IEA member crude oil and petroleum product inventory levels, refining capacity, and utilization rates.

Petroleum Products Markets Have Become Increasingly Global with Greater Trade and Prices Increasingly Linked across Countries

International trade in petroleum products has expanded significantly over the past 20 years, making the markets for gasoline, diesel, and jet fuel increasingly global in nature, and providing additional gasoline supply options for the United States. This trend has been particularly important for the United States, which has seen large increases in the volume of imported gasoline. A key impetus for global trade in petroleum products has been a structural surplus in production of gasoline and deficit in production of diesel in Europe as a result of a systematic switch in European countries to diesel-burning automobiles. While many experts we spoke with believe that growth in international trade of petroleum products will likely continue, they identified several factors that may limit or change the patterns of trade, including plans and mandates to introduce significant volumes of biofuels and the potential expansion of differing fuel specifications that a proliferation of biofuel blends would entail.

International Trade in Petroleum Products Has Expanded Significantly

International trade in petroleum products has expanded significantly over the past 20 years, making the markets for gasoline, diesel, and jet fuel increasingly global in nature. Specifically, our analysis of IEA data shows that OECD imports of gasoline, diesel, and jet fuel more than doubled between 1984 and 2006, from about 80 million barrels per month to over 160 million barrels per month. Similarly, OECD exports increased from about 55 million to over 140 million barrels per month over the same time period. While OECD exports and imports in these products have more than doubled, OECD demand for these products rose by less than 40 percent during the same time period. Figure 4 shows the increase in OECD imports and exports of gasoline, kerosene-type jet fuel, and diesel fuel.[9]

Trade in gasoline has been particularly important for the United States, which has seen large increases in the volume of imported gasoline. While in 1970 the United States was largely self-sufficient in gasoline, we now import over 10 percent of our annual consumption of gasoline and smaller percentages of jet fuel and some other products.[10] U.S. imports of gasoline and gasoline blending components, which accounted for about 31 percent of our imports of refined petroleum products in 2005, averaged about 1.1 million barrels per day, or more than 10 percent of U.S. daily consumption. According to DOE, imports have supplied about half of U.S. gasoline demand growth from 1993 to 2005.[11] These "total gasoline" imports include finished gasoline, which can be sold directly to retail markets, as well as gasoline blending components that are combined in the United States to make finished gasoline to serve various markets, some of which use special, cleaner-burning gasoline blends as part of their plans to meet federal air quality requirements.

A key impetus for global trade in petroleum products has been a structural surplus in production of gasoline and a deficit in production of diesel in Europe. This surplus of gasoline is largely the result of a systematic switch in European countries toward automobiles with diesel-powered engines, which are more efficient than gasoline-powered engines. European regulators promoted diesel fuel use in Europe by taxing diesel at a lower rate, and European demand for diesel fuel-powered vehicles rose. The European refining and marketing sector responded to this change in demand by importing increasing amounts of diesel, and exporting a growing surplus of gasoline to the United States. The United States has purchased increasing amounts of gasoline, including gasoline blendstocks, from Europe in recent years, as shown in figure 5. These imports have generally had a strong seasonal component, with higher levels of imports during the peak summer driving months and lower imports during the fall and winter. The major exception to this seasonality came in the months October 2005 through January 2006, when imports surged in response to U.S. shortfalls as a result of damage to and shutdowns of refineries and pipelines following Hurricanes Katrina and Rita in August and September 2005, respectively.

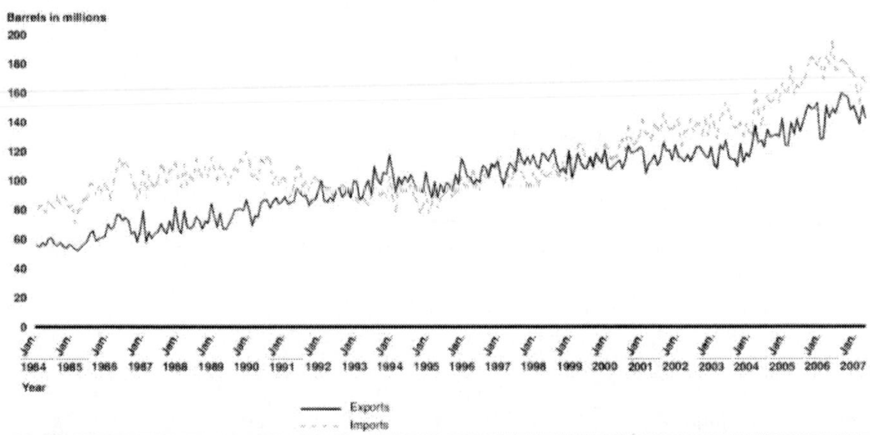

Source: GAO analysis of IEA data.

Figure 4. Imports and Exports of Gasoline, Kerosene-type Jet Fuel, and Diesel Fuel for All OECD Countries, 1984 - 2007.

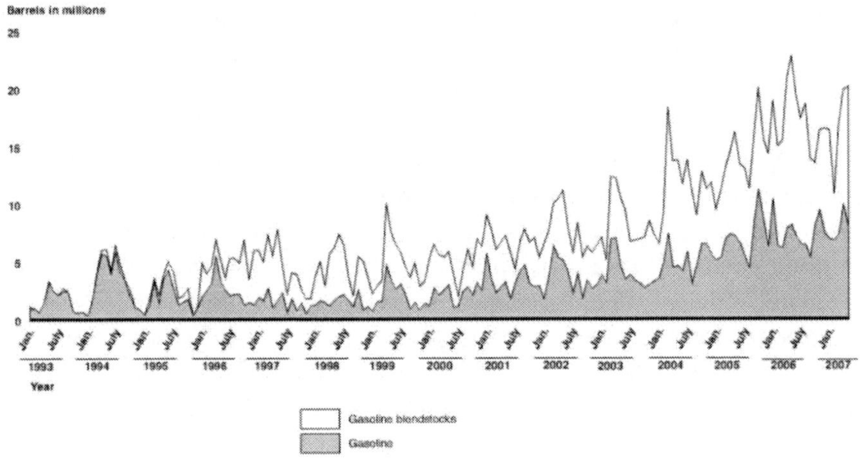

Source: GAO analysis of EIA data.

Figure 5. U.S. Imports of European Gasoline and Gasoline Blendstocks, 1993 – 2007.

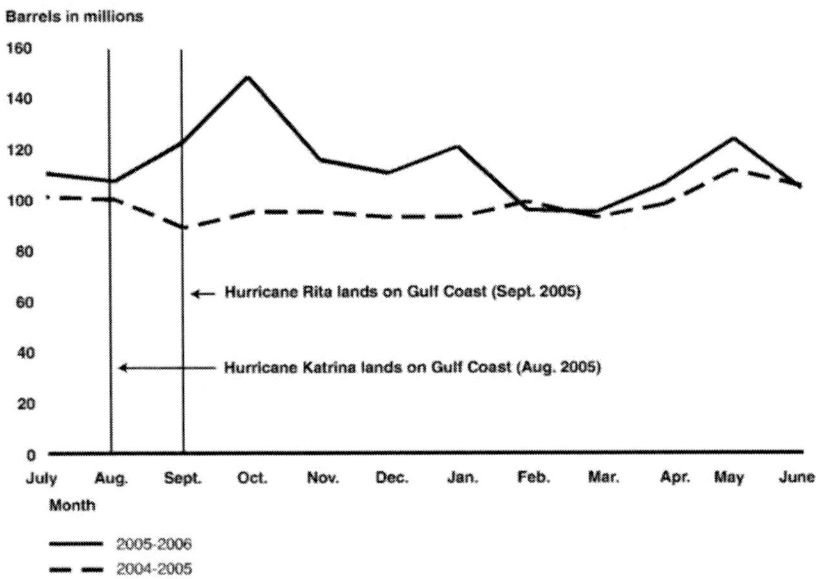

Source: GAO analysis of EIA data.

Figure 6. Petroleum Product Imports into the United States, 2004-2006.

Experts and company representatives told us they believe this structural imbalance within the European Union will continue for the foreseeable future, and perhaps widen, resulting in more exports of European gasoline and blending components to the United States. Specifically, company representatives and industry experts told us that European refiners are unlikely to significantly expand their refining capacity in the near future or reconfigure to produce less gasoline for a number of reasons, including the following:

- The profitability of the U.S. gasoline market acts as a draw for surplus gasoline worldwide. Many company representatives told us that the United States' continued appetite for gasoline—combined with many countries' declining demand—has resulted in most surplus gasoline being exported to the United States, and that this trend would likely continue in the future. For example, some refining interests in Europe told us they had configured their refinery operations to be essentially a U.S. "gasoline machine."
- Construction costs have increased significantly, raising the cost of investments in refining capacity or upgrades. For example, some refining interests in Europe and elsewhere told us that some planned conversion and upgrading of refinery capacity in Europe was on hold, because of increased construction costs worldwide. Some of these upgrade plans called for enhanced diesel fuel production mainly for the European market, as well as surplus gasoline exported to the United States.
- European refiners told us that they are reluctant to make large investments necessary to produce significantly more diesel because doing so will increase their greenhouse gas emissions. Their concern is that as greenhouse gas emissions caps are lowered, companies will be required to pay to reduce emissions or buy costly emissions credits.

EIA and other experts have stated that, at times, imports from Europe could be provided more competitively than gasoline from the U.S. Gulf Coast and other domestic refineries. In addition, more sources of supply can also benefit the United States in the event of domestic supply disruptions. For example, the flexibility in sources of supply helped U.S. marketers and retail sellers obtain gasoline and other petroleum products in the aftermath of Hurricanes Katrina and Rita in August and September of 2005, respectively, when a large fraction of the nation's refineries and pipelines were temporarily shut down. During the 3 months following the hurricanes, imports of gasoline to the United States increased by about 30 percent compared to what they were during the same months in the previous year, and imports came from a number of countries that do not typically sell to the U.S, market. Imports of other petroleum products into different regions of the country also rose. As illustrated by figure 6, U.S. imports of petroleum products surged in response to Hurricanes Katrina and Rita compared to levels during the same months of the previous year.[12]

In addition to gasoline, kerosene-type jet fuel imports into the Gulf Coast surged to about 3.3 million barrels in October of 2005, compared to just 20,000 barrels in October 2004. Some countries that did not export significant quantities of this fuel in 2004 exported significant quantities following the hurricanes to the United States in 2005. For example, France exported 580,000 barrels of kerosene-type jet fuel to the United States in October 2005, but nothing in October 2004 or October 2006.

Our analysis of wholesale prices in the United States, Europe, and Asia shows that prices in geographically dispersed markets rose significantly following Hurricanes Katrina and Rita, indicating that prices in these markets are linked to some extent. Because imports surged from many countries in response to the resulting supply disruptions in the United States, gasoline prices around the world rose along with prices in the United States before prices eventually returned to pre-hurricane levels. Figure 7 illustrates the price spikes that occurred in late August and late September 2005 as a result of the severe damage to oil and gas production facilities in the Gulf of Mexico and to refineries and pipelines onshore from Hurricanes Katrina and Rita. The figure clearly shows that European and, to a lesser extent, Asian spot gasoline prices—wholesale prices for gasoline traded on a daily basis at major market centers—responded to the resulting petroleum product supply disruptions in the United States. The additional supplies to U.S. markets from Europe and elsewhere reduced prices in the United States, and spot prices everywhere declined to pre-hurricane levels before the middle of October.

While experts have stated that the availability of additional sources of petroleum product supplies has benefited the United States through lower and less volatile prices, and foreign gasoline supplies clearly helped reduce prices following Hurricanes Katrina and Rita, the fact that petroleum product markets are international means the United States will be exposed to supply disruptions or unexpected increases in demand anywhere in the world. Further, because some foreign suppliers are further away from the U.S. demand centers they serve than the relevant domestic supply center, the length of time it takes to get additional product to a demand center experiencing a supply shortfall may be longer than had the United States had more refining capacity. For example, imports of gasoline to the West Coast may come from as far away as Asia or the Middle East, and the transport time and therefore cost is greater. To the extent that imported gasoline or other petroleum products come from far away, the lengthening of the supply chain has implications for the ability to respond rapidly to domestic supply shortfalls. Specifically, if supplies to relieve a domestic regional supply shortfall must come from further away, the price increases associated with such shortfalls may be greater and/or last longer. In this sense, the West Coast is more vulnerable to price increases or volatility than is the Northeast, which can receive shipments of gasoline into New York Harbor or elsewhere in the U.S. Northeast from Europe, often on voyages of less than a week.

Growth in International Trade of Petroleum Products Is Expected to Continue but Growth in Biofuel use May Limit or Change the Patterns of Trade

With demand for petroleum products growing globally, experts we spoke with believe the trade in petroleum products will continue to increase for a number of reasons. For example, global trends toward lower-sulfur fuels have resulted in more uniform sulfur specifications, creating more trade opportunities. Strong global demand for certain petroleum products—especially distillates such as diesel and jet fuel—will increase competition for, and facilitate global trade of, these petroleum products. For example, since 2005, diesel wholesale prices have generally been at a premium compared to the price of gasoline, in response to sharp consumer demand, and in the United States, diesel demand grew 6.9 percent in 2005, compared to 2.5 percent for gasoline. Demand for jet fuel is growing with the increase in air transportation, and given that jet fuel has uniform global specifications, jet fuel will continue to trade relatively freely based on global price signals.

While many experts we spoke with believe that growth in international trade of petroleum products will likely continue, the planned expansion of the use of biofuels, such as ethanol made from corn or other crops, and biodiesel made from soybeans or other crops, in the United States and many other countries could reduce the growth of demand for petroleum products and thereby reduce the opportunity for trade.

At the U.S. federal level, the EPA administers the Renewable Fuel Standard Program, which went into effect in 2007 and requires most U.S. gasoline refiners, importers and blenders to sell a minimum portion of biofuels each year. Refiners can meet the standard by blending biofuels with conventional gasoline or diesel in various proportions. Plans and mandates in a number of countries and regions, including the United States to introduce larger volumes of biofuels, primarily as additives to gasoline or diesel, could displace demand for and trade in petroleum products. In addition and as discussed previously in this report, some of these biofuel policies mandate that all gasoline or diesel sold in an area be blended in specific proportions with biofuels, with differences across states in the timing and level of such blending. Still other states specify a certain proportion of biofuels to be blended but allow flexibility in how they are blended, thereby creating the potential for widely different biofuel blended fuels within even a single state—for example, the mandated biofuel requirement could be met by a uniform proportion of ethanol blended into every gallon of gasoline used in the state, or by using a small amount of E85 (fuel composed of 85 percent ethanol and 15 percent gasoline components) with ethanol blended into the rest of the gasoline, and any combination of blends and volumes that meets the overall requirements would also satisfy the mandate. States and localities have pursued such policies for a variety of reasons, including viewing biofuels as a means to boost farm economies by increasing demand for feedstock crops while also contributing to a cleaner environment.[13] However, the current absence of uniform standards for biofuels and varying plans by various countries and regions to blend different volumes of biofuels with petroleum-based gasoline and diesel could reduce the fungibility of these fuels and thereby reduce opportunities for trade.

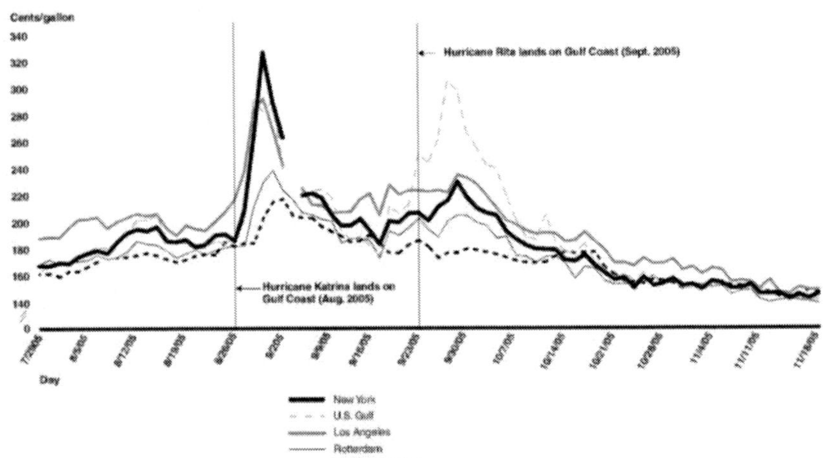

Source: GAO analysis of EIA data.

Note: Breaks in lines represent days for which no data were recorded.

Figure 7. Wholesale Gasoline Prices during the 2005 Hurricanes, 2005 – 2006.

Global and Domestic Refining Capacity Have Not Kept Pace with Demand, Leading to Tight Demand and Supply Balance and Recently Contributing to Higher Petroleum Product Prices

For most of the past 25 years, there has been excess refining capacity globally, but this excess capacity has been reduced over time as demand has increased faster than capacity has grown. Capacity growth has lagged behind demand for a number of reasons, including low profitability in the refining sector and demands on industry to meet changing fuel specifications and reduce emissions of environmental pollutants. More recently, unexpectedly rapid growth in demand for petroleum products caused refineries to run closer to their production capacity. Current market tightness has contributed to higher and more volatile prices and increased profits in the refining industry. While these higher profits have encouraged increased investments in refining capacity, it is unclear whether or for how long the current market tightness will continue. This uncertainty is, in part, because rising construction costs and uncertain future demand make it difficult to estimate how many of the planned refining projects will actually be completed and because biofuel initiatives in many countries could reduce demand for petroleum products while potentially requiring further refining investment to make and keep separate different gasoline and diesel specifications to be blended with ethanol and biodiesel.

Demand for Petroleum Products Has Grown More Quickly than Has Refinery Capacity, Tightening the Supply and Demand Balance Worldwide

For much of the past 25 years, demand for petroleum products in the United States and internationally has outpaced growth in refining capacity. Demand for petroleum products fell dramatically from 1978 to 1982, creating significant excess capacity—by 1983, almost 30 percent of U.S. and world refining capacity was idle. Demand for petroleum products began growing again around 1982, and this demand growth, along with the shutting down of some idle refining capacity, began to narrow the gap between capacity and demand. Since that time, growth in demand for petroleum products has generally exceeded growth in refinery capacity, causing refineries to run more intensively to meet demand. Figure 8 shows how refinery utilization in the United States and internationally, with a few exceptions, including the countries of the former Soviet Union, has increased significantly since the early 1980s.

Refining capacity in the United States has been growing since 1994 through expansions at existing refineries. The last major complex refinery on a new, or "green field" site in the United States was built in the 1970s, and many, mostly smaller, refineries were shut down starting in the early 1980s. However, as figure 9 shows, even as the number of refineries in the United States fell since 1981, refiners have since 1994 generally expanded total capacity at remaining facilities. Capacity expanded by an annual average of 192,000 barrels per day between 1994 and 2006—more than the average-sized refinery in 2006, which had a capacity of 116,000 barrels per day. For example, ExxonMobil's Baytown refinery grew by about 166,000 barrels per day in capacity between 1994 and 2006, more than equivalent to adding a new refinery. In this sense, it is potentially misleading to say that no new refineries have been built in the United States since the 1970s. Instead, experts have said that expansion in the United States has centered at existing facilities because such expansion is less expensive than building an entirely new refinery at a new, "green field" site because of lower construction, permitting, and land acquisition costs. Some industry officials we spoke with said that construction at a green field site can be about two to three times more expensive than expanding capacity at existing sites on a per barrel basis.

Figure 10 shows how much of the recent growth in refining capacity in the United States has been concentrated in the Gulf Coast. This growth in capacity in the Gulf Coast is consistent with the view of many industry experts we spoke with that the Gulf Coast provides one of the most competitive environments for U.S. refiners. Experts cited several factors, including ready access to imported crude oil supplies, numerous options for shipping product to the rest of the United States by pipeline and waterways, and a concentration of highly skilled workers.

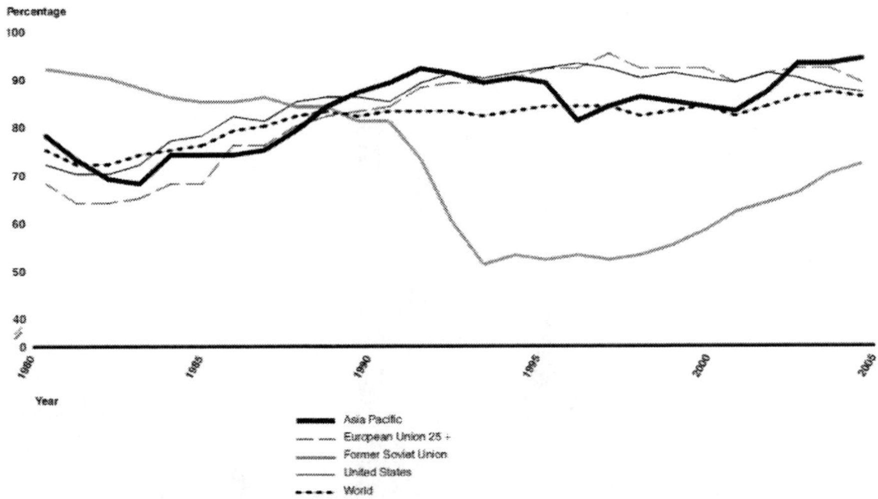

Source: GAO analysis of BP Statistical Review of World Energy June 2007.

Figure 8. Refinery Utilization in the World and Selected Countries, 1980-2006.

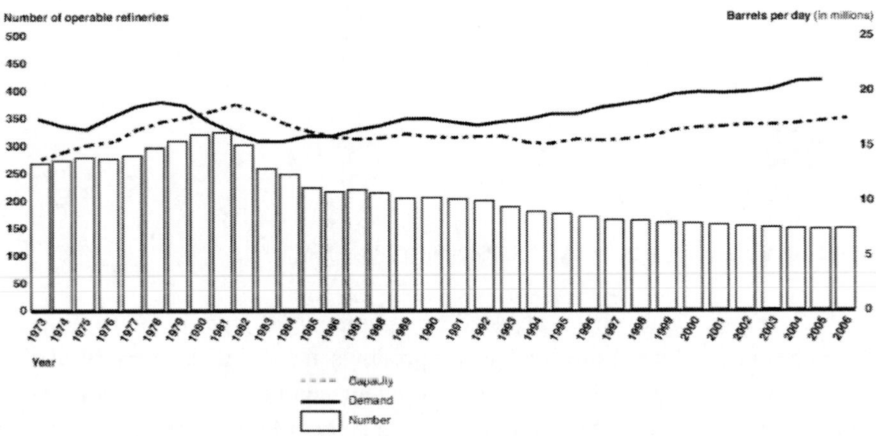

Source: GAO analysis of EIA data.

Figure 9. U.S. Crude Oil and Petroleum Product Consumption and Number and Capacity of Operable Refineries, 1973 - 2006.

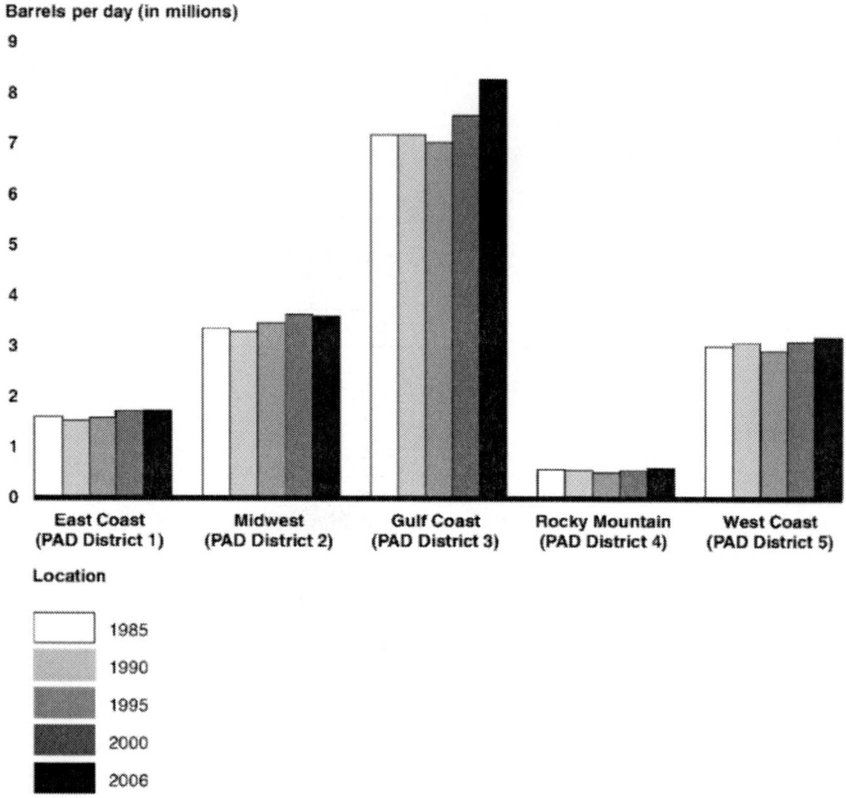

Source: GAO analysis of EIA data.

Figure 10: Refinery Capacity by U.S. Region, 1985-2006.

U.S. refineries also have invested in equipment to upgrade their refineries to be able to produce more high-value products from a wider variety of raw inputs. For example, hydrocracking equipment enables refiners to adjust the yields of various products, and coking capacity allows refiners to process heavier crude oils. Figure 11 shows how the capacity of such downstream units, particularly hydrocracking and coking, has grown faster than distillation capacity overall. The addition of such downstream units does not increase the distillation capacity of refineries—the traditional measure of capacity—but enables refineries to produce a greater portion of products in high demand (such as gasoline, diesel, and jet fuel) and also to process more heavy and sour crude inputs. In fact, the proportion of gasoline, diesel fuel and jet fuel produced per barrel of crude input in the United States has increased from 77 percent in 1993 to 81 percent in 2005 even as the quality of crude oil inputs used has deteriorated.

Industry officials and experts we spoke with said that several factors have caused refinery capacity to grow more slowly than demand in the United States. First, industry officials and experts said that refining has been a low-, even negative-return business for much of the past two decades, with profits too low to encourage significant expansion. Companies in the oil industry overall, which includes upstream oil exploration and production activities as well as downstream refining and retail marketing, have in general performed better than some industries and worse than others. However, according to an analysis by Deutsche Bank, cash returns on investment for oil companies in the Standard and Poor's 500 index were less than the cost of capital from 1986 to 2000. In other words, it cost companies more to raise the money to invest than those investments earned. Within the oil industry, the refining segment has been less profitable than other lines of business in the petroleum industry, according to EIA data, as illustrated in figure 12. Except for a few years since 1977, returns for U.S. refining and marketing operations have been lower than returns in foreign refining and marketing and lower than exploration and production. Specifically, during the entire period 1977-2005, average return on investment for the U.S. refining industry was 7 percent, compared to 9 percent for foreign refining and slightly over 10 percent for all other lines of business. For the integrated oil companies that still control a major portion of the nation's refining capacity, U.S. refining must compete with foreign refining and upstream investment options for capital. The lower returns for U.S. refining can make it harder for companies to justify expanding U.S. refining capacity.

Another indication that the refinery industry has long had low expectations of profitability is that existing refinery capacity has sold very cheaply. U.S. refineries have often sold for significantly less than what it would cost to build a new similar refinery. According to an analysis by the National Petroleum Council of the value of existing refinery purchases between 1998 and 2004, refineries sold for about one-fourth to one-third the cost of equivalent new construction. The cost of buying an existing refinery was also less than the general cost of expanding capacity on an existing refining site, which experts indicated could be less than one-half the cost of new construction. This suggests that refiners have had low expectations of future returns in the U.S. refining market. This also indicates that until recently, a refiner looking to expand capacity in the United States may have been able to do so more affordably by purchasing an existing refinery. This would add to that refiner's capacity, but would not expand domestic refinery capacity overall.

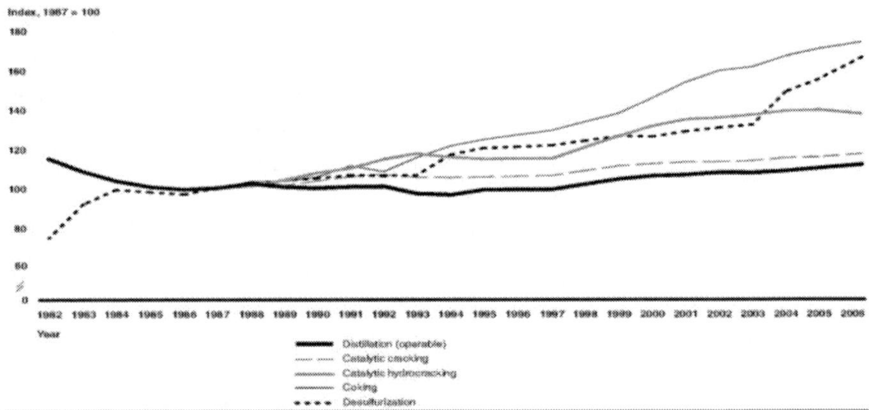

Source: GAO analysis of EIA data.

Figure 11. U.S. Refinery Distillation Capacity and Capacity of Selected Downstream Units, 1982-2006.

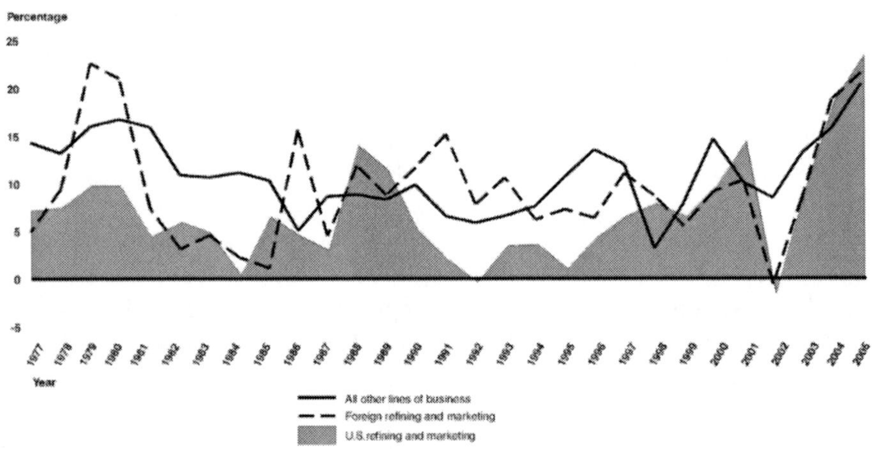

Source: GAO analysis of EIA data.

Note: This graphic is based on the performance of major energy producers covered by EIA's Financial Reporting System (FRS). These companies represented about 81 percent of U.S. refining capacity in 2005.

Figure 12. Return on Investment in U.S. and Foreign Refining Compared to Other Lines of Business for Major Energy Producers, 1997 - 2005.

A second reason experts cited for slow domestic refinery capacity growth is that more rigorous product specifications; the proliferation of special gasoline blends, or "boutique fuels" around the country; and environmental controls have all required refineries to invest in additional processes in order to meet the specifications and regulations, and these investments did not typically add to base capacity. Officials we spoke with said that the large investments required to reduce harmful air emissions at refineries and meet more stringent product specifications drew from the capital that may otherwise have been available to invest in expanding capacity.

A third reason for slow domestic refinery growth, according to some industry representatives, is that permitting difficulties have discouraged refinery expansions. Refineries are required to obtain permits from relevant state and local authorities in order to build or expand refinery capacity. These are often difficult to obtain owing to regulatory hurdles and public opposition. Other experts suggested that permitting adds to the difficulty of expanding capacity but has been a less important factor than overall low expectations of returns. DOE officials told us that resistance to refinery expansions by nearby communities could be a more important factor in discouraging new domestic refinery construction or expansions.

Finally, EIA officials and others pointed to the surplus production of gasoline in Europe as a major, more recent, reason domestic refinery capacity has not kept up with domestic demand. They stated that Europe could at times provide gasoline to the Northeast more competitively than some Gulf Coast refiners, and therefore gasoline imported from Europe has displaced domestic supplies and discouraged domestic refinery expansions.

More recently, global demand for gasoline, diesel, and jet fuel grew particularly quickly around 2004, partly as a result of rapid growth in China, where demand surged by over 15 percent. In order to increase production and meet this recent surge in demand for petroleum products, refiners have had to run their refineries even more intensively—capacity cannot be added quickly because of the long lead times involved in designing and constructing a refinery or an expansion project. Since 2004, world refinery utilization rates have risen to around 86 percent, the highest levels since 1980, when data first became available. Experts told us that many refineries have been running near their production capacity in order to meet demand. This is particularly true in the United States and Europe, where refineries have been running at or near 90 percent utilization since the 1990s, even though spare capacity still existed worldwide, particularly in the countries of the former Soviet Union and to a lesser extent in the Asia Pacific region.

Current Market Tightness Has Contributed to Higher Petroleum Product Prices, Higher Price Volatility, and Higher Industry Profits

The recent tightening of the balance between supply and demand for petroleum products has, along with higher crude oil prices and other factors, contributed to increased petroleum product prices and higher industry profits, and has contributed to greater price volatility. In addition, a tight demand and supply balance means less flexibility in industry's response to unanticipated events. For example, at times of excess capacity, if a particular refinery were to unexpectedly shut down for emergency maintenance, capacity that wasn't being used could be brought on line to meet demand. However, when refineries are generally running near capacity, there is less excess capacity to call on, and what available capacity there is tends to be located farther away from demand because the lower-cost and nearer refining capacity tends to be used up first.

An analysis by the FTC illustrated the effects of tight refining capacity at the regional level. This analysis compared the impact on gasoline prices of two refinery outages in the upper Midwest in the spring and summer of 2001 with a major refinery outage in Oklahoma in July 2003. Each of the Midwest refinery failures was associated with wholesale gasoline price increases of between 30 and 40 cents.[14] By contrast, the Oklahoma refinery failure was found to have little effect on gasoline prices in that state. The FTC attributed the difference in the price responses to the fact that the upper Midwest region lacks sufficient refinery capacity to meet the region's demand, while Oklahoma produces significantly more petroleum products than the state needs. Therefore, when a major disruption occurs, the upper Midwest must rely on supplies from distant refineries, while Oklahoma simply exports fewer petroleum products to other states.

Further, as we have previously reported, the West Coast of the United States generally has higher gasoline prices than much of the rest of the country.[15] Among the reasons for these consistently higher gasoline prices are a tight supply and demand balance for gasoline, the fact that the region is isolated from other major domestic and foreign refining regions, and the adoption in California of a unique blend of gasoline that is more costly to make than many other blends and that is not routinely produced by many refineries outside the West Coast. Prices in the West Coast can rise rapidly in response to supply disruptions as a result of these factors.

Profits in the refining industry have increased significantly since 2002, in part as a result of sustained market tightness and, in the United States, owing to wide price differentials between heavy and light crude oils. For example, the difference between crude oil input prices and petroleum product prices, a strong indicator of refining profits, has increased worldwide, though particularly in the United States. In the United States, these wide price spreads have caused returns on investments in the refining and marketing segment of the petroleum industry to reach record levels in 2004 and 2005, the latest data we were able to obtain. In the United States, these higher margins are, in part, the result of the ability of U.S. refineries to take advantage of low-price, low-quality crude oil inputs. Sophisticated U.S. refineries are able to convert large quantities of low- quality crude oil inputs into higher-valued products, while refineries in the rest of the world do not have such capacity to the same extent. Shifts in global crude oil production and demand have contributed to a glut of such low-quality oils, lowering their price relative to higher-quality crude oils and improving the position of U.S. refineries relative to that of their international competitors.

Increased Profit Margins Have Led to More Investment, but Future Market Tightness Will Depend on Several Factors

Currently high petroleum product prices and high profits in the refining industry have spurred new refinery capacity investments in the United States and internationally. Global investment in refining has increased in recent years. According to IEA data, capital spending grew from $34 billion in 2000 to $51 billion in 2005 and is expected to average $60 billion per year between 2006 and 2010. Analysis by IEA of plans and projects currently underway worldwide suggest that almost 10.5 million barrels per day of capacity may be added through 2011. This rate of refinery capacity growth is somewhat higher than expected demand growth, which is projected to grow by about 9.2 million barrels per day by 2011. The majority of this capacity expansion is expected to take place overseas, especially in China, India, and the Middle East. In the United States, EIA officials have said that announced investments through 2011 could add an additional 1 million barrels per day to domestic refinery distillation capacity, along with significant additions to downstream processing capacity. If realized, these domestic and international investments could help alleviate the tight balance between refining supply and demand. However, since tight refining capacity is just one of a number of factors affecting prices, the biggest factor being crude oil prices, even a less tight refining market may not bring much price relief at the gasoline pump.

While recent profits and prices have renewed interest in expanding refining capacity, experts said it is unclear whether or for how long current refining market tightness will continue. Future refining market tightness depends on changes in refining capacity and on changes in the demand for petroleum products. Industry officials and experts we spoke with said future conditions are highly uncertain for several reasons:

- It has become much more costly to expand refinery capacity in recent years due to rapidly rising construction costs. Various construction materials such as iron, steel, and concrete are important in energy projects, and their costs have increased significantly. For example, while prices for iron and steel fell in the decades prior to 2002, prices increased by 9 percent annually between 2002 and 2004, and by 31 percent from 2004 to 2005. Similarly, industry officials said that many decades of low investment levels have led to a small pool of qualified project engineers to design and oversee construction and expansion projects, causing labor prices to soar. Moreover, the Nelson-Farrar refinery construction cost index, which tracks prices for materials such as iron and steel, equipment and skilled and unskilled labor, shows that costs for refinery investment rose by 17 percent from 2002 to 2005 in real terms. Industry officials indicated that these cost estimates did not capture the full extent to which refinery expansion costs have increased. Officials also said that the waiting lists to purchase key refinery equipment are getting longer. In the United States and in Europe, some planned refining expansions have been delayed or canceled, in part because of these rising costs and delays in acquiring equipment and skilled labor.
- Uncertainty about future demand makes refinery investments risky and may inhibit investments. The United States is considered a mature market, with demand for motor gasoline forecast by EIA to grow by 1.2 percent annually between 2005 and 2030. Meanwhile, refinery capacity has on average expanded by almost 1 percent annually since 1999. Some industry officials we spoke with believe that U.S. demand for petroleum products will peak in the next decade and then begin to decline, implying only a temporary need for additional refining capacity. Company representatives told us that various proposals

under consideration dampen the U.S. refining investment climate. For example, the Administration has proposed to reduce U.S. petroleum gasoline consumption by 20 percent by 2017 through increased use of biofuels and more stringent automobile fuel economy standards. If achieved, this could turn the United States from a gasoline importer to a net exporter within 10 years; and current refining capacity could meet future demand even without expansions that are currently planned.

- Similar initiatives to blend large volumes of biofuels into the transportation fuels markets in other countries could similarly displace demand for petroleum products. It is unclear whether such initiatives could ease the demand and supply tightness that currently exists. On the one hand, reducing demand growth can reduce pressure on refinery capacity. On the other hand, reduced expectations of future demand can alter the attractiveness of refinery investments, and some refiners may respond by cutting back refinery expansion plans.
- New initiatives to blend biofuels in varying proportions into transportation fuels could potentially add to the need for further refining investments both to refine and to keep separate new blending stocks, possibly absorbing resources that could have been used to expand capacity. Automobile industry experts we spoke with agreed that each different ethanol blend requires a specific gasoline or diesel blend stock in order for the resulting blended fuel to meet performance and emissions standards. In other words, the gasoline that is blended with ethanol to make E10 (10 percent ethanol) is different than the gasoline used to make E85 (85 percent ethanol). The absence of national standards for blending biofuels with gasoline and diesel could also increase the number of gasoline and diesel blending stocks refiners have to make, and could lead to a proliferation of new blendstocks. Further, to the extent that new equipment is needed at refineries in order to produce, handle, or keep separate these various blendstocks, refineries will need to invest in this equipment in order to meet various federal, state, and local biofuel mandates and standards. These added investments could crowd out resources that could otherwise have gone to expanding refinery capacity.

Domestic and OECD Inventories of Petroleum Products and Crude Oil Have Declined Relative to Demand, with Mixed Effects on Prices and Price Volatility

When measured as average days of consumption, long-term trends in inventories of petroleum products and crude oil in the United States indicate a general decline over the past 20 years. Similarly, gasoline and crude oil inventories in OECD countries excluding the United States have also generally fallen over the same period. However, there are limitations to inventory data as measured by EIA and IEA, in part because these data do not fully match stocks with their intended markets; in general petroleum product exporting regions will typically have large stocks of these products relative to that region's demand, while inventories held in net importing regions will typically be lower relative to demand. For example, petroleum products stocks of gasoline in Canada, Europe, and the Caribbean that are destined for the United States are counted as inventories in those countries but not as inventories in the United States. A number of factors have contributed to the long-term decrease in inventory holdings in the United States, including reductions in domestic crude oil production and in the number of refineries as well as advances in technology and management processes that allowed for reduced inventories and a concomitant reduction in operating costs. Lower operating costs associated with lower inventories may have translated into lower consumer prices during normal periods. However, in the short term, inventory levels tend to fluctuate within a "normal" range, and—since inventories provide a smoothing effect against temporary demand and supply fluctuations—lower than normal inventories can lead to higher prices in the event of supply disruptions or surges in demand.

Inventory Levels of Petroleum Products in the United States and Other OECD Countries have Generally Fallen over the Past Two Decades

Privately held inventories of petroleum products and crude oil in the United States have generally fallen over the past two decades, in terms of average number of days worth of supply, or "days forward cover."[16] Specifically, as illustrated in figure 13, days forward cover for gasoline in the United States averaged about 30 days in 1984 but fell to an average of about 12 days for the first 5 months of 2007. Similarly, crude oil days forward cover fell from about 29 to about 22 days, and jet fuel and diesel fuel days forward cover also fell over the same period.

Other OECD countries have also generally seen a reduction in days forward cover for motor gasoline. Figure 14 shows the general downward trends in gasoline days forward cover for the OECD regions of Europe, Pacific, and North America excluding the United States. Specifically, European stocks declined from about 50 days in 1984 to about 40 days in 2001, before increasing to almost 46 days on average for the first 5 months of 2007. The much larger inventory figure for Europe compared to that for the United States reflects the fact that the inventory data include strategic stocks of gasoline held by some private companies.[17] The recent increase in European stocks of gasoline coincides with a period in which demand for gasoline fell in Europe relative to supply, and exports of gasoline to the United States increased a great deal. Figure 14 also shows large reductions in gasoline stocks in North America, excluding the United States, from about 44 to 18 days forward cover over the same time period, while stocks in Pacific OECD countries fell more modestly from 20 to almost 16 days.

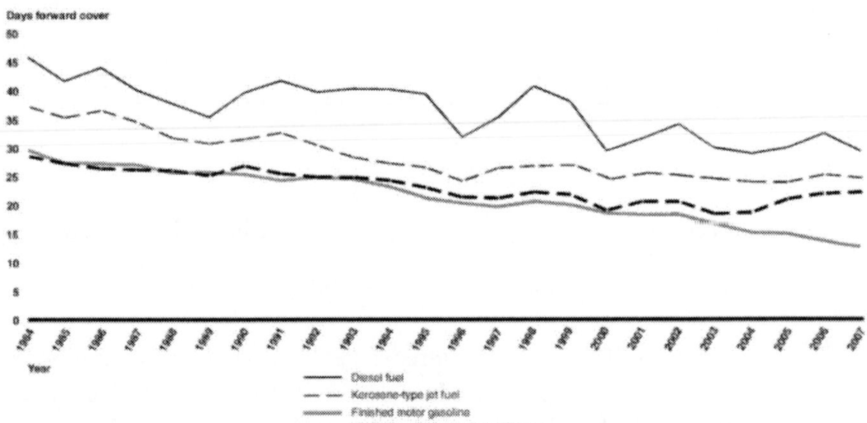

Source: GAO analysis of EIA data (annual average of monthly data).

Figure 13. Crude Oil (Excluding Strategic Reserves), Finished Motor Gasoline, Kerosene-type Jet Fuel and Diesel Fuel Days Forward Cover in the United States, 1984 – 2007.

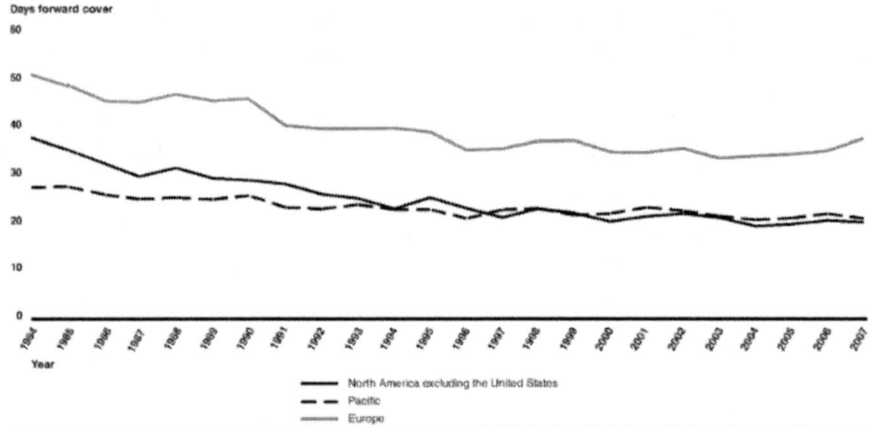

Source: GAO analysis of IEA data (annual average of quarterly and monthly data).

Figure 14. Motor Gasoline Days Forward Cover, by OECD Region, 1984-2007.

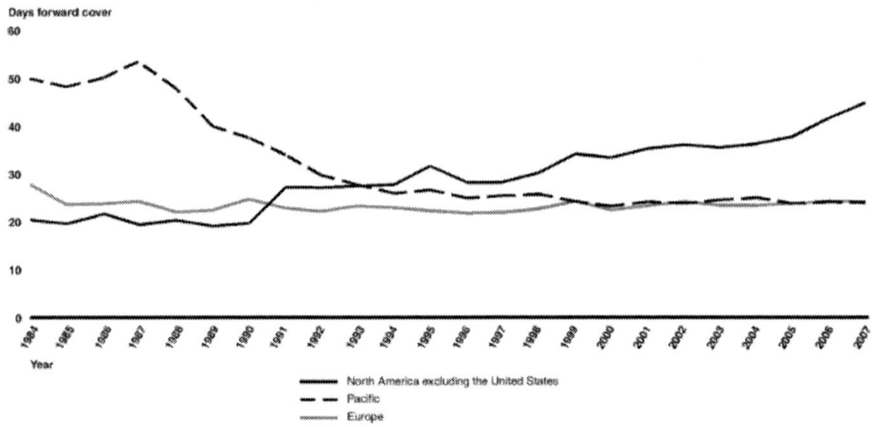

Source: GAO analysis of IEA data (annual average of quarterly and monthly data).

Figure 15. Crude Oil Days Forward Cover, by OECD Region, 1984 - 2007.

Crude oil stocks in two of the three other OECD regions, Europe and Pacific, decreased over the period, while stocks in North America excluding the United States rose significantly, driven in part by increases in Canadian oil sands production and the storage and delivery infrastructure associated with this increased production. Figure 15 illustrates these changes in crude oil inventories in days forward cover. We do not have inventory data for non-

OECD countries. However, as with petroleum products, net crude oil exporting countries would be expected to have much higher levels of days forward cover for crude oil than net importing countries, especially if strategic stocks are excluded.

Inventories, as measured by IEA, EIA, and others, have some limitations as a measure of what is available to meet demand in the event of a supply shortfall. For example, as discussed above, the United States has imported an increasing share of its gasoline over the period during which inventories have fallen, and as such, the domestic inventory data do not account for large volumes of these products on the water in tankers from foreign sources that are destined for the U.S. market or in storage terminals at foreign ports serving this trade in gasoline. Our analysis indicates that about 16 million barrels of gasoline and gasoline-blending components were en route to the United States on the average day during 2006, representing about an additional 2 days of forward cover, and an unknown additional amount is held in storage terminals that would be available for shipment in the event of a supply shortfall in the United States. Data on U.S. gasoline inventories may further under-represent available inventories if we ignore the part of inventories held at foreign refineries that are intended to serve the U.S. petroleum products market. The inventories represented in these refineries' storage systems and in the tanker and pipeline system supporting the flow of products to the United States, or at least some portion, could be considered part of U.S. inventories for the purpose of evaluating our days forward cover of products in the event of a supply disruption. However, it would be difficult to estimate the precise volumes of these foreign inventories as we do not collect such data from offshore suppliers and because many of these inventories are at varying distances from U.S. markets and would have to be evaluated differently, depending on how long they would take to reach the United States in the event of a domestic supply shortfall.[18] Another limitation in interpreting inventory data arises because much of the measured volumes of petroleum products in pipelines cannot be effectively removed from the pipelines in the event of a supply shortfall because they are needed to keep the pressure in the pipelines at operable levels. Similarly, some inventories are in so-called "tank bottoms," or the part of storage tanks that cannot effectively be retrieved in the event of a supply shortfall. As a result of these and other limitations, we do not have an accurate measure of precisely how much is in the full supply chain to the United States, or the actual number of days' worth of usable supply we could rely on in the event of a supply disruption.

Looking forward, the refining expansions discussed previously in this report may lead to increases in the days forward cover measure as pipelines and storage facilities associated with the new refining capacity add to inventory holdings. However, any increase in days forward cover is likely to be modest overall because demand is also projected to grow, and companies continue to strive to develop more efficient inventory holding practices and reduce costs associated with holding any excess inventory.

A Number of Factors Have Contributed to the Long-Term Decrease in Inventory Levels Since 1980

A number of factors have contributed to the long-term decrease in U.S. days forward cover. These factors include (1) a reduction in the number of refineries and falling domestic crude oil production, (2) the fact that demand has been rising faster than refining capacity for much of the past 20 years, (3) gains in technological and management efficiency that have allowed companies to reduce the level of operating inventories, and (4) the rise of futures markets for crude oil and gasoline that have enabled oil companies and others to reduce exposure to market risk by holding financial as well as physical stocks of these commodities.[19]

In the United States the decline in U.S. crude oil production resulted in decreased inventory in gathering pipelines and storage infrastructure, as pipelines and storage tanks were decommissioned. This decline in production-related inventories could be quite significant, although we do not have data to measure it directly. Oil production in the United States peaked in 1970 at around 10 million barrels per day, but by 2005 had fallen to less than 6 million barrels per day. This decline in production has left a number of abandoned crude oil pipelines and therefore represents a reduction in measured inventories. Similarly, the closure of many small refineries and the decommissioning of these refineries' storage and pipeline connections to the greater supply infrastructure also reduced inventories held at these facilities and in the pipeline connections. As discussed previously in this report, this reduction in the number of refineries was significant. For example, in 1980, there were well over 300 refineries in the United States, while in 2006 the number was about 150. This sheer drop in numbers probably overstates the drop in associated inventories because, while the number of refineries fell, the average size of refineries rose, both because it was smaller refineries that tended to be shut down and because many of the remaining refineries expanded their capacity significantly. Nonetheless, EIA has stated that refinery closures had an important impact on petroleum stocks prior to 1995.

A related cause of the reduction in days forward cover for petroleum products has been the fact that for much of the past 20 years, demand for these products has risen faster than domestic refining capacity. Because days forward cover is measured as the number of days' worth of demand that is in the domestic supply chain, any increase in demand that is not met by a commensurate increase in domestic supply will lead to a reduction in days forward cover. However, as discussed previously in this report, this is potentially misleading because the supply chain between foreign refiners and the United States is relevant for measuring actual days forward cover and the U.S. imports of gasoline have been increasing dramatically over the past decade, and imports of crude oil have been increasing for much longer than that.

According to company representatives and industry experts we spoke with, as well as the National Petroleum Council, delivery system efficiency improvements have also resulted in reduced crude oil inventory levels. Company representatives told us improved information technology has given managers better tools needed to optimize stock levels, and that this was mirrored in many other industries over this same period, as improved logistics and management practices enabled companies to more closely track production and delivery. By reducing inventories, refiners were able to reduce their operating costs, providing incentives to invest in efficiency- improving measures. Some officials told us that low refining profit margins were a major driver in getting companies to reduce their inventory holdings.

Finally, the relationship between the future price of crude oil and petroleum products and the amount of inventory stored has, at times, contributed to changes in overall inventory levels. For example, according to a 1997 EIA report, during the period between 1995 and 1996, the prevalence of instances where the current trading price of crude oil or petroleum products was higher than the future expected price appeared to be an important factor behind the reduction in stocks.[20] In such a price environment, oil companies and others could sell currently held inventories and buy futures contracts to meet their future expected needs. In recent years, the future price of crude oil has most frequently been higher than the current price, and industry officials told us that this has been a factor in explaining why absolute inventory levels increased between December 2004 and early 2007. For example, total U.S. crude oil stocks, excluding strategic stocks, were about 355 million barrels in

June 2007, or about 31 percent higher than in January 2004, according to EIA data.

Long-Term Inventory Cost Reductions have Likely Reduced Prices of Gasoline and Other Petroleum Products, but, In the Short Term, Reductions in Inventory Levels below Normal Ranges Can Lead to Higher Prices during Supply Shortfalls

To the extent that improved technology and inventory management over the long term have resulted in lower operating costs, some of the savings may have been passed on to consumers in the form of lower prices. We found no consensus among industry experts about the extent of such price reductions, nor any empirical analyses that would quantify the savings to consumers from lower inventory holding costs. However, because refineries compete with one another to sell their products, they would likely be forced to pass on some savings in operating costs in order to remain in operation, especially during that portion of the last three decades in which there was unused refining capacity as well as during recent years when surplus gasoline production in Europe has increasingly found its way to U.S. markets.

However, because inventories provide a smoothing effect against temporary demand and supply fluctuations, lower than normal inventories can signal underlying changes in supply and demand conditions that will cause prices to rise. For example, if a large refinery in the United States were to suffer an unexpected outage, the resulting reduction in domestic supplies would likely result in a drawdown of that refinery's inventories to meet its demand, and if that is insufficient, the refinery would buy from other refiners. If inventories were on the high end of the normal range, such a disruption would likely have little effect on petroleum product prices, all else remaining the same. On the other hand, if inventories were on the low side of or below the normal range—the result of other supply shortfalls or unexpectedly high demand—the additional refinery outage would be more likely to cause significant price increases. The size of the supply disruption relative to available inventories, as well as to the size of the refining sector, can also influence how prices respond. For example, if a large refinery outage were to occur in the Gulf Coast refining region, the large volume of inventories and the large number and capacity of other refiners relative to that refinery's production would likely mean that the effect on prices of petroleum products would be small. Similarly, the availability of large stocks of gasoline in

Europe, often less than 1 week away by tanker to the U.S. East Coast market, probably insulates the latter market from extreme price fluctuations in the event of a domestic supply shortfall. On the other hand, if the refinery outage were to occur in the West Coast, where one refinery's production would be significantly larger relative to available inventories and total sector capacity, a significant price response is more likely. For these reasons, lower than normal inventories are frequently cited as a factor in price run-ups of petroleum products. For example, the Northeast diesel price spike of January 2000 was preceded by lower than normal inventory buildup, and the California Energy Commission cited relatively low inventory levels as one of several contributing factors to the gasoline price spike in the spring of 2006.

U.S. Supply Infrastructure Is Constrained in Key Areas and Likely to Become Increasingly Constrained, Thereby Increasing Prices and Price Volatility unless Timely Investments Are Made

The nation's crude oil and petroleum product supply infrastructure is constrained in key areas and may become inadequate to handle future volumes of petroleum products and biofuels unless sufficient investment is undertaken. Inadequate supply infrastructure can lead to higher prices and price volatility during supply disruptions or unexpected increases in demand because the supply infrastructure cannot handle the changed or increased delivery of fuels. However, the extent of the problem and the prospect for the future of the supply infrastructure is uncertain, in part because there has been no comprehensive study done to assess infrastructure adequacy. There are many planned infrastructure expansions that could alleviate the stress on the system to some extent. However, a complex approval process—involving numerous federal, state, and private entities—and other factors increase the time and cost of building and maintaining infrastructure.

The Nation's Supply Infrastructure Is Constrained in Key Areas and Likely to Become More Constrained

Industry and agency officials report that key crude and petroleum product pipelines are constrained and operating at or near capacity. As the Secretary of

Energy noted in a December 4, 2007 discussion with industry media, the U.S. energy infrastructure system—including oil pipelines—is "pressed," and it is important that pipeline and other energy infrastructure owners maintain their assets effectively, to maintain adequate supplies. Both DOT and industry officials report a systemic lack of pipeline capacity in the supply infrastructure system in key states including Arizona, California, Colorado, and Nevada, and note existing pipeline supply infrastructure is insufficient to carry the commensurate volumes of petroleum products and crude oil needed to meet growing demand. Industry officials told us that pipelines in the Southwestern region, such as Arizona and Las Vegas, have reached maximum utilization, or become "constrained." For example, industry experts told us that a new petroleum product pipeline from the Gulf of Mexico to El Paso is already approaching full capacity. Denver's petroleum product pipelines have also become generally constrained and unable to meet increased gasoline demand for summer travel. This raises the cost of delivering petroleum products to Denver; in instances when pipelines are full, shippers must make alternate shipping arrangements by more costly rail or truck. Further, a key petroleum product line from the U.S. Gulf Coast to North Carolina is reportedly constrained, thereby increasing delivery costs for petroleum products in that region, and key petroleum product pipelines radiating outward from the major refining center of Houston are also reportedly constrained. Finally, in certain areas, pipeline infrastructure to support certain demand or production centers' needs does not exist. For example, there are no petroleum product pipelines into Florida. Additionally, despite strong demand in California, the existing petroleum product pipelines support the flow of product from California to other Southwestern states, but no petroleum pipelines flow into California from other regions.

Industry representatives and federal studies also report that many of the nation's port facilities are operating at or near capacity. For example, one-fourth of the ports in a U.S. Maritime Administration (MARAD) survey described their infrastructure impediments as "severe." Officials from the interagency U. S. Committee on the Maritime Transportation System, which includes MARAD, the National Oceanic and Atmospheric Administration, and the U.S. Army Corps of Engineers told us that U.S. ports and waterways are constrained in capacity and utilization, and anticipate marine supply infrastructure will become more constrained in the future. The Ports of Los Angeles, Long Beach, Oakland, Houston, Savannah, and Charleston reported congestion and emphasized in a 2005 report that they are experiencing higher than projected growth levels.

The capacity of the supply infrastructure not keeping pace with increasing demand in certain areas has raised concerns about the adequacy of the infrastructure to accommodate expected increasing volumes of crude oil and petroleum products. Population increases in the West and South are expected to increase the need for pipelines, marine transportation, and capacity utilization there. DOT reports that already high pipeline capacity utilization levels may not meet growing demands unless significant expansion occurs. The situation is similar for the U.S. marine infrastructure. In a 2005 report, MARAD evaluated the status of U.S. ports and waterways and concluded that domestic marine transport supply infrastructure will become more constrained in the future. As imports of petroleum products are projected to increase by over 80 percent by volume between 2004 and 2030, according to EIA, this anticipated demand growth will challenge a marine transport system that is already operating, in some instances, at the limits of its capacity.

The introduction of biofuels will also increase the strain on the existing supply infrastructure. For example, ethanol-producing plants tend to be relatively small near the sources of biofeedstocks—currently mostly corn—used to make ethanol. At present, the ethanol produced by these plants, unless they are located next to rail facilities, are typically trucked to central rail loading facilities and then shipped to demand regions on unit trains—trains whose cars are entirely made up of a single product and typically going to a single destination. Trucking biofuels to these central locations is costly and also uses petroleum products, thereby reducing the volumes of these latter fuels the ethanol can displace. Experts we spoke with generally agreed that eventually a more efficient collection system will likely be built—probably consisting of feeder pipelines—to connect the relatively small ethanol plants to major rail or supply and demand centers. Nonetheless, according to DOE, the existing petroleum product pipelines are currently not configured to transport ethanol from regions where it is currently produced to regions where it is consumed. Because pipelines are ultimately the cheapest form of domestic shipment of petroleum products and crude oil, it may make sense to ultimately ship ethanol through the pipeline system, and existing or new petroleum pipelines could be used in certain areas to transport ethanol if ongoing efforts by operators to identify ways to modify their systems to make them compatible with ethanol or ethanol-blended gasoline are successful.

In addition, as discussed previously in this report, a proliferation of biofuel blends in this country will require additional variations in the blends of petroleum products that are mixed with these biofuels. Pipeline companies report that varying fuel specifications complicate petroleum product delivery and supply infrastructure systems by requiring separate storage and increasing the complexity of the distribution system. Also, pipeline operators told us that sending more and smaller batches of these special blends has slowed the flow of fuels through pipelines because pulling off more and smaller batches of fuels requires a slower speed to not miss significant parts of these batches. However, when we asked, these pipeline operators did not offer any quantification of the extent to which effective tank capacity reduction or pipeline slowness has occurred.

Infrastructure Disruptions Lead to Increases in Prices and Price Volatility and Constraints in Supply Infrastructure Could Exacerbate Price Effects

A constrained supply infrastructure can be a major factor influencing prices of petroleum products during supply disruptions. For example, during the rupture in the Kinder Morgan pipeline in Arizona in August 2003, Arizona's gasoline prices rose by about 45 cents during the 3-week period ending on August 25, 2003. Due to the connectivity of the pipeline network among California, Arizona, and Nevada, the disruption not only caused prices to spike in Arizona itself, but the extra burden from Arizona's demand also contributed to higher prices elsewhere in the West; during the disruption, California's prices rose by 40 cents to peak at $2.10, and Washington, Nevada and Oregon all experienced price increases of over 30 cents per gallon.[21] Any constraint in the supply infrastructure can reduce supply reliability by making it more difficult to reallocate supplies in response to even relatively minor disruptions in the supply and distribution system. In this way, a constrained supply infrastructure could increase price volatility and exacerbate price effects due to disruptions.

When certain localities are inadequately served by pipelines or reasonably priced marine supply infrastructure, alternative transport modes tend to be more costly, leading to higher prices for consumers. For example, since relatively few pipelines connect the West Coast with other regions, some supplies of petroleum products and crude oil must be shipped by truck or barge from other domestic regions or by tanker from foreign countries; such modes of transport are slower and more costly than via pipelines. For example, it can take around 2 weeks for a vessel to travel from the Gulf Coast to Los

Angeles port—including transit time through the Panama Canal. This can increase recovery time from an unplanned refinery outage, other supply disruption, or an unanticipated surge in demand, thereby leading to higher or longer-lasting price spikes.

Federal agency officials and industry experts told us that the slow permitting process and corresponding delays in infrastructure development could lead to higher and more volatile petroleum product prices in the future. For example, while the recent expansion of pipeline capacity from the Gulf Coast to El Paso, following the opening of the Longhorn pipeline in June 2004, has been expected to ease the infrastructure constraint on Arizona's petroleum product supplies, permitting impediments continue to perpetuate the lag between the growth of demand for petroleum products on the West Coast on the one hand and the growth of the pipeline capacity to move products to the region on the other. The California Energy Commission has recently stated that similar constraints on marine infrastructure expansions to accommodate future growth in demand for imports of petroleum products will be a major challenge for the West Coast. Such failure of the region's supply infrastructure to handle the requisite volumes of petroleum products to meet rising demand will continue to contribute to the persistence of higher and more volatile prices in the West Coast compared to other regions.

We were unable to assess the extent of supply infrastructure constraints or the impacts of these constraints on prices and price volatility, in large part because there is no central source of data that tracks system bottlenecks. Information that would indicate whether a pipeline is operating at or near capacity is also not collected in a central location by federal agencies or industry trade groups. These data would include pipeline throughputs, measured by the amount of product flowing into a pipeline and the volume of output received at key market locations. Companies are not required to report such information. By contrast, FERC requires natural gas pipelines to report, via their web sites, throughput information that allows regulatory, public, and private entities to track bottlenecks and identify where shortages in supply, or system constraints, affect regional prices. A number of studies and analyses of constraints in natural gas pipelines have quantified the effects on natural gas prices. For example, EIA routinely uses natural gas pipeline capacity and outages in making projections about natural gas prices. These data on natural gas pipeline capacity and the flow of natural gas are collected and evaluated to determine the reliability of the infrastructure to meet demand, and it is well understood that constrained pipelines lead to higher natural gas prices and can even lead to disruptions of service in severely constrained cases.[22]

We recognize there are differences between the natural gas industry on one hand and the petroleum industry on the other, particularly because of the fact that the former industry evolved under a rate-regulated utility framework, while the petroleum industry did not. Specifically, under rate regulation, the former requirement that utilities meet all demand at their regulated prices at any point in time necessitated the monitoring of supply and infrastructure constraints that could cause a failure of service. By contrast, petroleum product prices have largely not been regulated, and prices have generally been allowed to adjust to equilibrate supply with demand at any point in time. Further, we are not suggesting in this report that petroleum product markets should be regulated like natural gas or any other markets. However, these historical regulatory differences notwithstanding, we believe that it is important to understand the extent to which constraints on the current petroleum product supply infrastructure affect prices as well as the adequacy of the infrastructure to meet growing demand. Federal agencies, industry experts, and Congress have all recognized this as a priority. For example, industry consultants and agency officials have acknowledged the importance of a system-wide study of pipeline capacity constraints and regulatory impediments to future investment. In addition, DOT officials have stated that the extent of capacity restrictions in the nation's pipeline infrastructure is becoming more apparent, that the current regulatory mechanisms may not lead to appropriate reinvestment in the industry. In June 2006, DOT put forth a proposal and in December 2006 Congress passed legislation that mandated the Secretaries of Energy and Transportation to conduct periodic analyses of the adequacy of the nation's pipeline supply infrastructure. The first report to Congress of the results of such an analysis is required by June 2008.[23] The language for the mandate stated that "such analyses should identify areas of the United States where unplanned loss of individual pipeline facilities may cause shortages of petroleum products or price disruptions and where shortages of pipeline capacity and reliability concerns may have or are anticipated to contribute to shortages of petroleum products or price disruptions. Upon identifying such areas, the Secretaries may determine if the current level of regulation is sufficient to minimize the potential for unplanned losses of pipeline capacity." Despite widespread recognition that such a study is needed to fully identify the extent of infrastructure inadequacy and the impact on prices, to date, no such analysis has been undertaken. DOT and DOE officials told us that they were not appropriated funds specifically to do the mandated analyses and that the agencies have not re-allocated other funds for this, although DOE told us in its comments that DOE and DOT staff have met to discuss how this work could be approached. Given that the study has not begun, it seems highly unlikely that the agencies will be able to meet their June 2008 deadline for reporting to Congress.

Expansions in Supply Infrastructure Are Planned, but High Construction Costs, Investment Risk, and a Complex Regulatory Environment Can Deter or Delay These Needed Infrastructure Investments

There are many private sector plans to expand the supply infrastructure, and if implemented in timely fashion, these plans could significantly alleviate the stresses on the system. For example, there is a long- anticipated project for a 500-mile petroleum product pipeline expansion from Louisiana to Georgia, several plans for new crude pipelines to accommodate the expected increased flows of Canadian oil sands, as well as other crude and refined product pipeline plans to meet more localized needs. However, many such plans are in a conceptual stage and/or subject to permitting approval and other possible complications. Thus, industry representatives told us, it is difficult to determine how many of the industry plans for new construction or expansion of existing pipelines will be realized.

However, the high cost of construction, uncertain investment climate, and complex regulatory environment increase the time it takes to build this supply infrastructure and raises risk and investment costs. With regard to construction costs, a shortage of skilled labor and specialized equipment to perform the work, and high prices of steel and concrete have increased construction costs and the time it takes to expand the nation's supply infrastructure system. For example, pipeline companies and other industry experts we spoke with said that major pipeline expansion and construction projects take anywhere from 2 to 15 years to complete and currently cost about $1 million per mile to build. With regard to the uncertain investment climate, pipeline companies and industry experts told us that uncertainty about petroleum product demand, biofuel development and shipping, and future changes to fuel specifications complicate the decisions about where and when to build new or expand existing infrastructure. Regulations governing pipeline and other infrastructure expansions, including regulations governing water and air pollution, endanger

ed species protection, and public safety, have evolved to protect the environment and ensure public safety. However, there can be tension between these goals and the goals of ensuring adequate energy supplies and keeping prices down. For example, in order to build a new pipeline or significantly expand capacity or upgrade an existing pipeline, companies must first navigate a mixed and sometimes complex jurisdiction of federal, state, and local regulators, as well as secure right of way approval from the necessary landowners whose lands will be crossed by the pipeline. At the federal level alone, as many as 11 agencies may be involved in granting approval to build new pipeline projects. In addition, industry experts told us that some potential market entrants have had difficulty meeting permitting requirements and are often unable or unwilling to wait out lengthy delays in obtaining permits, such as when two companies in southern California reportedly recently backed out of plans to build storage terminals there after trying to complete the federal, state and local approval processes. A study conducted for Association of Oil Pipe Lines, an FTC report on gasoline prices, and industry officials told us that building or expanding pipelines has become increasingly difficult in certain situations. For example, a major pipeline operator encountered federal and local legal and regulatory issues that delayed for 10 years the development of a key pipeline from the Gulf Coast to El Paso, Texas. As a result of such delays and impediments to investment, regional demand that could support new pipeline capacity must be served by more costly transportation modes for years, as has been the case in parts of the Rocky Mountains and Southwest and West Coast regions. Finally, an uneven balance of costs and benefits of expansion for various entities can also contribute to declining investment in supply infrastructure by certain entities. For example, DOT reports that common carrier pipelines achieve only modest returns from relieving constrained pipeline capacity. However, it reports consumers would benefit proportionately greater through the enhanced competition resulting from the increased capacity of new pipeline investments. Pipeline companies, on the other hand, report they will expand when sufficient demand is secured, particularly through the "presale" of capacity in the proposed pipeline.

Ideally, the permitting and approval process should be streamlined without sacrificing the important protections provided by regulatory oversight. Industry and federal agency officials have pointed out that a federal model exists for this in the permitting process for interstate natural gas pipelines. Specifically, FERC facilitates expansions and construction of natural gas pipelines by serving as the lead agency to process company permit applications, conduct the required environmental impact study, and coordinate the timing of other necessary permits that fall under the purview of various federal agencies. In addition, FERC authorizations convey the right of eminent domain to pipeline builders to resolve specific right of way issues in the event an agreement cannot be reached between a landowner and a project sponsor. FERC officials told us that although its authorizations convey the right of eminent domain, pipeline companies rarely have to exercise it because its existence is usually sufficient to get landowners to negotiate a solution with pipeline builders. Streamlining the federal regulatory process with regard to crude oil and petroleum product pipeline repairs has already begun in response to a federal statute passed in 2002 to coordinate environmental reviews and permitting needed for pipeline repairs and more clearly define federal roles in the pipeline repair process.[24] However, this streamlined federal process has not been applied to constructing new crude oil or petroleum product pipelines or significantly upgrading or increasing capacity of existing pipelines.[25]

CONCLUSION

The choices the United States and other countries make about how to ensure sufficient supplies and stable prices of petroleum products and other fuels such as ethanol and biofuels will greatly influence energy prices in the United States. For biofuels in particular, cost and availability will depend in part on how well international, federal, state, and local governments coordinate their biofuel standards and methods of integrating them with petroleum products. Harmonizing fuel specifications worldwide, while continuing to allow for regional differences in fuels specifications that are there to meet specific environmental or vehicle performance goals, would make it easier to refine and transport common blends, streamline delivery, increase opportunities for trade, provide additional sources of supply, and potentially reduce prices and price volatility. However, if the world and the United States end up with numerous different biofuel blends—as appears to be happening under existing plans and mandates—this could expand the array of incompatible gasoline and diesel blending stocks and final blended products that cannot be interchanged at the retail level, reducing opportunities for trade. In addition, these products will have to be segregated during shipment, further straining the supply infrastructure. Unless the supply infrastructure catches up and keeps up with these changes, the domestic energy supply will be less secure and prices will tend to be higher or more volatile.

Even without these changes, rising demand for crude oil and petroleum products over the last 25 years has challenged the supply infrastructure for these commodities in certain areas, leading to higher prices during supply disruptions or during periods when pipelines or ports lack sufficient capacity to transport the products suppliers wish to ship. As noted by the Secretary of Energy in December 2007, the U.S. energy infrastructure system—including oil pipelines—is "pressed," and it is important that pipeline and other energy infrastructure owners maintain their assets effectively, in order to maintain adequate supplies of energy. In the absence of a comprehensive analysis of the likely weaknesses in our infrastructure, policy makers and regulatory agencies involved in overseeing the safety and adequacy of supply infrastructure remain in the dark about the extent of these problems and their effects on prices of petroleum products. Further, as demand for petroleum products and biofuels grows, the existing system may become increasingly constrained and need to be upgraded and expanded to handle greater and different product flows. Because federal and state agencies and other entities will be involved in approving such upgrades and expansions, it is essential that they be well informed as to the current state of the supply infrastructure and the areas in most critical need of further investment. Furthermore, the lack of a lead agency to streamline the complex and costly permitting process for U.S. supply infrastructure construction or expansion projects and the lack of ability of federal agencies to convey the power of eminent domain in cases where conflicts over infrastructure placement cannot be resolved may deter potential market entrants from investing in much- needed upgrades in a timely fashion. As a result, we could end up with less security of supply and higher and more volatile prices in the future.

RECOMMENDATIONS FOR EXECUTIVE ACTION

To better monitor and evaluate the development of our nation's supply infrastructure systems, as well as to facilitate the continued tradability of products across domestic and global markets and to ensure that gasoline supplies from Europe and elsewhere remain compatible with U.S. gasoline specifications, we are making a number of recommendations that, if adopted, should improve prospects for the future security of petroleum product supplies and price stability.

- To avoid additional proliferation of differing fuel specifications that would further burden the existing supply infrastructure and create impediments to trade, we recommend that the Secretary of Energy coordinate with EPA and other relevant federal agencies, states, IEA, the European Union, and other foreign entities to encourage development of biofuels and petroleum products standards and blending practices that maximize the fungibility of these fuels and minimize the spread of differing fuel types that would further strain the supply infrastructure, while recognizing that some fuel differences to reflect local environmental requirements, engine performance, or other factors are likely beneficial.

- To comprehensively analyze the U.S. supply infrastructure's capacity to accept, handle, and transport the increasing volumes and types of petroleum products and biofuels expected to traverse its system, we recommend the Secretaries of Energy and Transportation undertake the comprehensive study of existing and projected increases to the infrastructure system—including terminal capacity and pipeline throughputs—to evaluate whether future demand is likely to be met by existing infrastructure and planned increases as mandated by Congress in 2006. To the extent that the data to comprehensively conduct such analyses may at present not be collected, the Secretaries should consider evaluating the merits of enhancing the reporting of utilization and throughputs, perhaps using natural gas pipeline and storage reporting requirements as a model.
- In conjunction with the completion of the first comprehensive study of the supply infrastructure, we recommend the Secretary of Transportation work with DOE, FERC, EPA, and other federal agencies to evaluate the feasibility and desirability of designating a lead federal agency, with authority to convey the power of eminent domain, to coordinate across agencies and streamline the permitting and siting process for crude oil and petroleum product interstate pipeline expansions, upgrades, and new construction, using FERC's role with natural gas pipelines as a model. If this is found to be feasible and desirable, we recommend the aforementioned agencies work together to determine which agency should take the lead role and to prepare a legislative proposal for Congress to provide any additional authority needed to implement this recommendation.

Agency Comments and Our Evaluation

We provided the Departments of Energy and Transportation and the Federal Energy Regulatory Commission with a draft of this report for their review and comment.

DOE neither agreed nor disagreed with our report and primarily provided technical comments, which we incorporated as appropriate. However, we do note that the Secretary of Energy recently commented that the U.S. energy infrastructure system—including oil pipelines—is "pressed," and that it is important that infrastructure assets are maintained effectively, which appears to be in accord with our recommendations. Specifically, with regard to our recommendation that the Secretaries of Energy and Transportation undertake the comprehensive study of existing and projected increases to the infrastructure system, it would seem that such a study would be critical in determining the level of maintenance and development needed to maintain adequate supplies of crude and refined products, a matter of national interest, as stated by the Secretary of Energy. Further, in light of the Secretary's comments, it would seem prudent that DOE also implement our recommendation to work with other federal agencies to evaluate the feasibility and desirability of designating a lead federal agency, with authority to convey the power of eminent domain, to coordinate across agencies and streamline the permitting process for crude oil and petroleum product interstate pipeline expansions, upgrades, and new construction.

DOT gave us oral comments on the report. With regard to our second recommendation, DOT commented that the agency supports the idea of conducting the comprehensive infrastructure study, as mandated by Congress in December 2006—in fact, a DOT 2006 legislative proposal also included language about the need for such a study—although agency officials added that they need funding to accomplish this task. In its comments, DOT said it believed our report unfairly portrayed DOT's inaction with regard to starting this mandated study as a failure. Specifically, it said this was not fair because the agency was not appropriated the funding to do the study. In response to these comments, we added language to the report that points out that DOT had proposed such a study to Congress. We already had language in the report acknowledging that DOT and DOE had said that they were not appropriated funding for the study. Still, our recommendation is in accordance with DOT's 2006 legislative proposal to perform this study, and we believe DOT and DOE should take steps to begin this study, either by reallocating their current budget or, if this is not possible, to request additional funding from Congress. With regard to the third recommendation, to study the streamlining of the petroleum product pipeline permitting and siting process, DOT commented that the agency supported streamlining the process for expanding petroleum product pipelines and had already proposed legislative language to Congress in 2006 that would have done so and also would have made DOT the lead federal agency in coordinating this process. Therefore, DOT commented that it had already done what we are asking in our third recommendation: namely, to evaluate the feasibility and desirability of designating a lead federal agency to coordinate the federal permitting process for crude oil and petroleum product interstate pipeline expansions, upgrades, and new construction. In reviewing DOT's legislative proposal, it does not call for the lead agency to have the ability to convey the power of eminent domain, as we also recommend be considered. Further, we cannot, on the basis of our work, endorse a specific agency to take the lead. Finally, we think that a coordinated effort among DOT, DOE, and other relevant agencies is needed to evaluate this issue and advise Congress on the best way to proceed. For these reasons, our recommendation is still appropriately addressed to the Secretary of DOT and the other named agencies.

FERC generally agreed with our findings and recommendations in the draft report, and provided technical comments, which we addressed in the body of the report as appropriate. FERC's letter, underscores FERC's agreement with GAO's recommendations and points out that FERC's role as the lead agency for siting natural gas pipelines could serve as a good model to use in interagency discussions about how this could be accomplished in the case of petroleum product pipelines.

Mark Gaffigan
Acting Director
Natural Resources and Environment

APPENDIX I: SCOPE AND METHODOLOGY

The Chairman and a member of the Senate Commerce, Science, and Transportation Committee asked GAO to evaluate trends and effects on petroleum product prices in (1) international trade of petroleum products; (2) refining capacity and intensity of refining capacity use internationally and in the United States; (3) international and domestic crude oil and petroleum product inventories; and (4) domestic crude oil and petroleum product supply infrastructure, particularly pipelines and marine transportation.

To address the first objective, we examined data from the Department of Energy's (DOE) Energy Information Administration (EIA) and the International Energy Association (IEA) to evaluate trends in the international trade flows for crude oil and petroleum products and their price correlations over time at international trading hubs. In addition, IEA data were used to calculate total global imports and exports of crude oil and petroleum products as well as for key global regions including Europe, Asia and the United States. We met with more than 20 oil industry companies—including refiners and pipeline companies—a number of financial and investment corporations, more than 25 industry groups, and more than 15 domestic and international government agencies to corroborate trend analyses, reports, and data. We conducted audit work in various locations in Texas, California, New York, and Washington, D.C., as well as Belgium, France, Germany, and the United Kingdom to obtain industry's perspective on recent trends in the international trade of petroleum product as well as prospective trends going forward. In addition, we analyzed EIA and New York Mercantile Exchange, (NYMEX) data on historical spot and futures prices for crude oil and petroleum products at international and domestic trading hubs to see how price volatility has changed over time.

To address the second objective, we assessed trends in refining capacity, refining capacity additions, utilization, complexity, and planned investments using IEA, EIA, and *Oil and Gas Journal* data, and determined the data were sufficiently reliable for our purposes. We met with more than 20 oil-industry companies—including refiners and pipeline companies—a number of financial and investment corporations, more than 25 industry groups, and numerous staff and officials of more than 15 domestic and international government agencies in California, Texas, New York, Washington D.C., as well as Belgium, France, Germany and the United Kingdom to corroborate trend analyses, reports and data. We also reviewed and analyzed trends in refinery investment, operating costs, and profitability in the U.S. and internationally, using literature and data on U.S. and international refining practices, trends and forecasts, and interviewed experts on these trends.

To address the third objective, we used data from EIA and IEA on crude oil and petroleum product inventories and projected demand to conduct international, U.S. total domestic, and U.S. Petroleum Administration for Defense District (PADD) inventory trend analysis on inventories in absolute terms and in "days forward cover" terms. We analyzed NYMEX and other futures market data, as well as EIA data, to observe the effects of the expected future price for crude oil on inventory holding decisions. To collect these data, we conducted a site visit to meet with industry and government representatives in Belgium, France, Germany, and the United Kingdom to gain information about the European Union's policy of maintaining strategic petroleum product reserves and their effects on price levels and price volatility.

To address the fourth objective, we interviewed federal and state agencies that oversee the economic, safety, and environmental impacts of pipelines and marine transportation on current and future utilization capacity of the petroleum product infrastructure. Where possible, we collected and analyzed data on the age of the pipeline and marine infrastructure system, capacity, throughputs, and constraints. We compared data reporting requirements for petroleum products with reporting requirements for liquefied natural gas, and identified differences in such reporting requirements. We spoke with common carrier pipeline operators, port authorities, government entities, and trade association and consumer advocate groups to gain their perspectives on supply infrastructure investment, capacity utilization levels, and potential system constraints. We also reviewed previous relevant GAO reports and testimonies, and Department of Energy and Department of Transportation reports. In addition, we examined reports and data from supply disruption case studies to examine those cases' impact on infrastructure, prices, and price volatility.

During our audit work we consulted with the following entities:

- We met with the following oil industry companies, including refiners, supply infrastructure and oil service companies: BP; Buckeye Partners; Chevron Corporation; ConocoPhillips Company; ExxonMobil Corporation; Fluor Corporation; Frontier Oil Corporation; Hess Corporation; Holly Corporation; Kinder Morgan Energy Partners; Longhorn Partners Pipeline; Magellan Midstream Partners; Marathon Oil Company; Mid-continent Express Pipeline; Oiltanking GmbH; Paramount Petroleum Corporation; Plains All American Pipeline L.P.; RaceTrac Petroleum, Inc.; Sunoco, Inc.; TEPPCO Partners L.P.; Tesoro Corporation; UOP LLC; Valero Energy Corporation.
- We met with the following financial organizations: Deutsche Bank; Goldman, Sachs & Co; JP Morgan Chase Bank; Morgan Stanley; New York Mercantile Exchange, Inc. (NYMEX).
- We met with the following industry groups and expert institutions: Allegro Energy Consulting; American Association of Port Authorities; American Petroleum Institute (API); Association of Oil Pipe Lines (AOPL); Conservation of Clean Air and Water in Europe

(CONCAWE); Consumer Federation of America; Energy Analysts International, Inc.; European Petroleum Industry Association (EUROPIA); Global Insight, Inc.; Institut Francais du Petrole (IFP); Muse Stancil & Co.; National Association of Regulatory Utility Commissioners; National Petrochemical & Refiners Association; *Oil & Gas Journal*; Petroleum Marketers Association of America; Pipeline Safety Trust; PIRA Energy Group; Purvin & Gertz, Inc.; Stillwater Associates LLC; Turner, Mason & Company; the Rabinow Consortium, LLC; UK Petroleum Industry Association; Union of European Petroleum Independents (UPEI); University of California Energy Institute; Western States Petroleum Association; Wood Mackenzie Research and Consulting.

- With regard to government and agency sources, we met with the following U.S. agencies and governmental institutions: Department of Defense, including the Army Corps of Engineers; Department of Energy, including the Energy Information Administration; Department of State; Department of Transportation, including Pipeline and Hazardous Materials Safety Administration (PHMSA); Department of Homeland Security; Federal Energy Regulatory Commission; Federal Trade Commission; Interagency Committee on Marine Transportation; Oak Ridge National Laboratory. We met with the following state and local governmental agencies: California Energy Commission (CEC); California Environmental Protection Agency Air Resources Board (CARB); Hawaii Energy Planning and Policy Branch; Port of Houston Authority. We met with the following international government and multilateral organizations: European Commission Directorate-General for Energy and Transport; EBV (German Stockholding Agency); French General Directorate for Energy and Raw Materials; International Energy Agency (IEA); International Monetary Fund (IMF).

The report primarily uses data from the domestic and international wholesale petroleum product and crude oil markets. In contrast to retail markets, wholesale prices do not generally include extra costs such as federal and state taxes, distribution and marketing expenses and profits. In every case for the data used in this report, we assessed and determined that the data were sufficiently reliable for our purposes. We performed our work from August 2006 through September 2007 in accordance with generally accepted government auditing standards.

End Notes

[1] GAO, *Energy Markets: Mergers and Other Factors That Influence Gasoline Prices*, GAO-07-894T (Washington, D.C.: May 23, 2007); GAO, *Energy Markets: Factors Contributing to Higher Gasoline Prices*, GAO-06-412T (Washington, D.C.: Feb. 1, 2006); GAO, *Energy Markets: Gasoline Price Trends*, GAO-05-1047T (Washington, D.C.: Sept. 21, 2005); GAO, *Motor Fuels: Understanding the Factors That Influence the Retail Price of Gasoline*, GAO-05-525SP (Washington, D.C.: May 2005).

[2] FERC also serves as the lead agency in coordinating the permitting process across federal agencies and can similarly convey the right of eminent domain for electricity transmission lines.

[3] The Organisation for Economic Co-operation and Development is a group of 30 countries committed to democracy and the market economy to support sustainable economic growth, maintain financial stability, and assist other countries' economic development. These countries are Australia, Austria, Belgium, Canada, Czech Republic, Denmark, Finland, France, Germany, Greece, Hungary, Iceland, Ireland, Italy, Japan, Korea, Luxembourg, Mexico, Netherlands, New Zealand, Norway, Poland, Portugal, Slovak Republic, Spain, Sweden, Switzerland, Turkey, United Kingdom, and the United States. However, we consider the United States separately for the purposes of this report.

[4] GAO, *Gasoline Markets: Special Gasoline Blends Reduce Emissions and Improve Air Quality, but Complicate Supply and Contribute to Higher Prices*, GAO-05-421 (Washington, D.C.: June 2005).

[5] The Strategic Petroleum Reserve (SPR) is a federally maintained stockpile of about 700 million barrels of light crude oil for use in the case of a major disruption of oil supplies.

[6] Access to the rail market is limited and tanker trucks' expenses depend on distances traveled.

[7] By petroleum products, we refer to primarily gasoline, diesel, jet fuel, heating oil. Most petroleum products and crude oil are shipped primarily by pipeline within the United States. Imports of petroleum products and crude oil, however, travel to the United States mainly over sea by vessel.

[8] FERC also collects administrative, financial, and operational information on crude oil and petroleum product pipeline companies.

[9] Figures represent trade originating or ending in OECD member nations, including trade between OECD nations, from OECD nations to non-OECD nations, and from non-OECD nations to OECD nations. Because figures include some trade from OECD nations to other OECD nations, such trade is counted as both an import and an export and therefore includes some duplication of counting. Furthermore, figures do not account for trade between non-OECD nations and therefore understate the total global trade of these products.

[10] Total gasoline includes both finished motor gasoline and motor gasoline blending components.

[11] Imports of distillate fuels and jet fuel have also risen in the last 20 years, while imports of residual fuel oil have declined.

[12] The graphic shows that imports remained significantly higher than in the same months during the previous year at least through January 2006. This was likely the result of lasting damage to U.S. refining production caused by the hurricanes.

[13] States are subject to some federal requirements in setting biofuel policies. Currently, EPA has determined that only blends of up to 10 percent ethanol are allowed in conventional gasoline vehicles and blends of up to 85 percent ethanol are permitted in flexible fuel vehicles. However, the state of Minnesota and the Renewable Fuels Association, as well as DOE are developing research and tests to gather the data required to facilitate EPA certification of fuel blends up to E15 or E20.

[14] The price increases were measured in gasoline prices in Chicago relative to Houston prices.

[15] See, for example, GAO's *Motor Fuels: Gasoline Prices in the West Coast Market,* GAO-01-608T, (Washington, D.C.: Apr. 25, 2001).

[16] In the United States, inventory data reported in this report refer only to privately held stocks, not the federally held crude oil and heating oil strategic reserves. As will be discussed later in this report, this is not true of some other OECD member inventory data.

[17] Unlike in the United States, where the federal government holds strategic stocks of primarily crude oil—but also a relatively small stock of fuel oil in the U.S. Northeast Home Heating Oil Reserve—European countries hold a large fraction of their strategic stocks in petroleum products, including gasoline and certain distillate fuels. Some European countries require private companies to maintain these stocks.

[18] It is also not clear that the benefits of collecting and maintaining such data outweigh the costs. Evaluating these trade-offs was beyond the scope of this report, but such an evaluation would have to be made before making a decision to collect a broader range of inventory data.

[19] Assessing the relative importance of these factors with any precision would be very difficult and we did not undertake this task in this report, so the list of factors should not be seen as a ranking of those factors in any way.

[20] Energy Information Administration, Petroleum 1996: Issues and Trends, (Washington D.C., September 1997).

[21] Note that there may have been additional factors influencing prices during this period, so we are not asserting that the pipeline outage was responsible for the entire change in prices.

[22] It should be noted that whether or not the benefits of collecting and maintaining such data outweigh the costs is unknown. Evaluating these trade-offs was beyond the scope of this report, but such an evaluation would have to be made before making any decision to collect a broader range of pipeline or other infrastructure data.

[23] Pub. L. No. 109-468, §8.

[24] Pub. L. No. 107-355, § 16

[25] In 2006, DOT identified the need for additional Congressional authority to reduce the regulatory burden on companies trying to construct new pipelines or repair existing ones. Specifically, DOT proposed legislation that, according to DOT, would among other things provide "minimal authority" to assist pipeline operators in overcoming state and local-level impediments to constructing new pipelines and would further streamline the permitting process for pipeline repairs. At this time, Congress has not provided this additional authorization. DOT's proposal did not call for a federal agency to have the authority to convey the power of eminent domain in cases where conflicts over infrastructure placement cannot be resolved but it would have authorized the Secretary of DOT to "designate an ombudsman to assist resolving disagreements between Federal, State, and local agencies and pipeline operators arising during agency review of pipeline repairs and hazardous liquids pipeline construction projects…"

INDEX

D

E

F

G

H

I

J

P

Q

R